John Gmeiner

Modern Scientific Views and Christian Doctrines Compared

John Gmeiner

Modern Scientific Views and Christian Doctrines Compared

ISBN/EAN: 9783744659567

Printed in Europe, USA, Canada, Australia, Japan

Cover: Foto ©Lupo / pixelio.de

More available books at **www.hansebooks.com**

PREFACE.

A PROMINENT American unbeliever observed, not long ago, in a leading periodical: "A profound change has taken place in the world of thought. . . . Christians excuse themselves for belonging to the Church, by denying a part of the creed. The idea is abroad that they who know the most of Nature believe the least about Theology. The sciences are regarded as infidels, and facts as scoffers."

It is, alas! but too true that many intelligent people of our country have become either open opponents or indifferent to the truths of Christianity, because they labor under the false impression, that Modern Science has brought facts to light which demolish the very foundation of Christian belief. It is taken for granted by many so-called advanced thinkers, that the Mosaic Account of the Creation and primeval history of mankind has been completely refuted by the results of modern scientific research, that no vestige of a Creator or Supreme Ruler of the Universe is to be found in visible Nature, that the soul of man is essentially alike to the life-giving principle of brutes, and its immortality, at least, doubtful, etc.

Such and similar infidel views, which are destructive of the very fundamental doctrines of Christianity, are more or less openly taught in many schools, colleges, and universities, and published in books, pamphlets, and papers, which are circulated throughout the country. Hence, we need not be surprised to find that many of our intelligent reading people, among them lawyers, physicians, teachers, editors students, and even mechanics and farmers, have become imbued with would-be scientific views incompatible with Christian belief. In fact, no Christian family-circle is secure from the baneful influence of infidel doctrines.

To ignore the disagreeable fact, that the tendency of many modern so-called advanced thinkers is decidedly infidel and anti-Christian, will not remedy the evil nor neutralize its influence on the Christian youth of this country. The only effective way of combating infidelity, in the usurped garb of Modern Science, seems to be, to show that all assertions about conflicts between Religion and Science are perfectly unfounded.

This I have endeavored to do in the following pages. Believing modern would-be scientific infidelity to be, probably, the most insidious and formidable foe the Church has to contend with in this country, a favorite study of mine for the last fifteen years has been, to compare the results of modern scientific research with the doctrines of Christianity.

While preparing the following pages for the press, it has been my aim to avoid all one-sided or narrow-minded views, knowing these to be the fruitful sources of the so-called conflicts between Science and Religion. For this reason I have consulted on the different points treated, the works of eminent scientific authorities of various shades of religious belief, Catholic, Protestant, indifferent, and infidel, as the reader will notice in glancing over the names of the authors mentioned. I have, moreover, endeavored to select as far as possible, only really eminent and reliable scientific and theological authorities, for such alone can fairly be considered representative exponents of Modern Science and Religion. To avoid the danger of misrepresenting the statements of these authorities, I deemed it best, as a rule, to give their exact words, as is usually done in publications wherein the views of different writers are compared; for the exact meaning of an author can seldom be given any better than in his own words.

I take pleasure in expressing here my sincere thanks to the learned reverend gentlemen, William Mahoney, of St. John's Cathedral, at Milwaukee, and S. Lebl, D. D., Professor of Philosophy, Sacred Scripture, etc., in our Seminary, for their kind and valuable assistance in preparing these pages for the press.

May this humble volume be welcome, and, with God's blessing, useful to many: to such as live in infidel society, or occasionally associate with unbelievers; to such as have infidel friends to whom they would wish to give a suitable book to read; to such as sometimes read infidel publications; to students of seminaries, colleges, and universities, who cannot ignore modern scientific views nor their bearings on Christian doctrines; and to reverend pastors in whose congregations infidel influences may be undermining Christian religion.

The sentiments with which I offer this book to the reader, I cannot better express than in the words of the great teacher St. Augustine (De Vera Religione, 20): "*Quae vera esse perspexeris, tene et Ecclesiae Catholicae tribue; quae falsa, respue et mihi, qui homo sum, ignosce.*"

REV. JOHN GMEINER,

St. Francis, near Milwaukee, Wis.

July 16, 1884.

ERRATA: Page 15, line 12, last word, read "all." instead of "alone." Page 64, line 17, fourth word, read "dry" instead of "day."

CONTENTS.

I. INTRODUCTORY REMARKS.

THE TWO BOOKS OF GOD.

THERE are two different books by which God teaches mankind—the Book of Nature and the Book of Revelation. Even Thomas Paine observes:[1] "The universe is the Bible of a true Theophilanthropist. It is there that he reads of God * * * Contemplating the universe, the whole system of creation, in this point of light, we shall discover that all that which is called natural philosophy is properly a divine study. It is the study of God through His works * * * Do we want to contemplate His power ? We see it in the immensity of the creation. Do we want to contemplate His wisdom ? We see it in the unchangeable order by which the incomprehensible Whole is governed. Do we want to contemplate His munificence ? We see it in the abundance with which He fills the Earth. Do we want to contemplate His mercy ? We see it in His not withholding that abundance even from the unthankful * * * It has been the error of the schools to teach astronomy, and all the other sciences, and subjects of natural philosophy, as accomplishments only; whereas they should be taught theologically, or with reference to the Being who is the author of them : for all the principles of science are of divine origin."

The other book by which God teaches mankind, is His

---

1. The Theological Works of Thomas Paine, Chicago, 1882, pp. 200-1.

supernatural Revelation, which He made by persons inspired or guided by His Spirit, and by His Incarnate Son. The means which God has given to man, to read these two different books, are—Reason, to read the Book of Nature, and Faith, to read the Book of Revelation. No doubt, *different* truths are taught in the Book of Nature, and in the Book of Revelation; and many of the truths taught in these two books, seem to be little related to one another. For instance, Geometry has little to do with the Doctrine of Justification,—Geology, little with Baptism; Botany, and Astronomy, seem to be perfect strangers to the Doctrines of the Incarnation, Trinity, etc.

We may say, these two books contain two quite different realms of truth; the one, the truths of the natural order, which man can investigate with the aid of his own natural faculties; and, the other, especially "mysteries hidden in God, which, unless divinely revealed, cannot be known by men." At first sight, it might seem that Nature and Revelation, Science and Christian Belief, are mutually independent and perfectly unrelated to one another. Yet, as Church and State, though quite distinct, now and then meet on some few points : so, also, some of the truths of Nature, or Science, and some Doctrines of Revelation, occasionally come in contact with one another; and it is on some such points, unbelievers claim, that there exists a conflict between Science and Revelation.

Now, God being the source of all truth, the Author of both Reason and Revelation, it is impossible that He should teach anything as true in His Revelation, and contradict the same in His works of Nature. Therefore, the Church has always held that there never can exist any real contradiction, or conflict, between Nature and Revelation, between Reason and true Christian Belief.

The late Vatican Council declared: "Although Faith is above Reason, there can never be any real discrepancy between Faith and Reason, since the same God who reveals mysteries and infuses Faith, has bestowed the light of Reason on the human mind, and God cannot deny Himself, nor can truth ever contradict truth. The false appearance of such a contradiction is mainly due either to the dogmas of faith not having been understood and expounded according to the mind of the Church, or to the inventions of opinion having been taken for the verdicts of Reason. *We define, therefore, that every assertion contrary to a truth of enlightened Faith is utterly false* * * * And not only can Faith and Reason never be opposed to one another, but they are of mutual aid one to another; for right Reason demonstrates the foundations of Faith, and enlightened by its light, cultivates the science of things divine; while Faith frees and guards Reason from errors, and furnishes it with manifold knowledge. So far, therefore, is the Church from opposing the cultivation of Human Arts and Sciences, that it in many ways helps and promotes it * * * Nor does the Church forbid that each of these Sciences, in its sphere, should make use of its own principles and its own method; but, while recognizing this just liberty, she (the Church) stands watchfully on guard lest Sciences, setting themselves against the Divine

Teaching, or transgressing their own limits, should invade and disturb the domain of Faith."

These words express the doctrine the Church has always held as to the relations between Science and Divine Revelation. This doctrine may be expressed in these few words : *"No truth of Science does, or ever can, contradict any truth of Divine Revelation."*

ASSERTIONS OF MODERN UNBELIEVERS.

Modern unbelievers, of course, assert that Science has refuted doctrines of Revelation. Mr. Ingersoll, whose boldness in making unfounded assertions is well-known,[2] declares in his lecture on Thomas Paine : "Since his (Thomas Paine's) day, it (the Bible) has been proven false in its Cosmogony, false in Astronomy, false in its Chronology, false in its History, and so far as the Old Testament is concerned, false in almost everything."

It is strange that only infidels can see that any doctrines of Divine Revelation have been refuted by Science. Vast numbers of learned Christians, who have studied more carefully than ordinary infidels, both the doctrines of Revelation and the results of modern scientific investigations, cannot see any conflict whatever between Science and Religion.

How is it to be explained that our infidels chronically imagine to have discovered some such conflict? The only plausible explanation that can be given, is contained in the famous words of Alexander Pope : "A little learning is a dangerous thing ;" for, as the English philosopher Bacon observes : "A little philosophy

2. See Notes on Ingersoll, by Rev. L. A. Lambert, Buffalo, N. Y.

inclineth man's mind to Atheism, but depth in philosophy bringeth men's minds about to religion."—Our infidels have learned just enough to see some seeming difficulties between Science and Revelation, but, unfortunately, not enough to be able to solve these difficulties.

It is a great delusion of our modern infidels, to imagine that their infidelity is the product of Modern Science. Modern infidelity had its origin with the so-called Deists of England, in the 17th century,—long before the various branches of so-called Modern Science, as Geology, Paleontology, Spectrum-Analysis, etc., were much heard of. Another great delusion of some of our infidels is, to imagine that all great scientists were infidels. Facts prove the contrary. Copernicus, Newton, Kepler, Euler, Heinrich Steffens, Von Schubert, Chas. von Raumer, John von Fuchs, Andrew and Rudolf Wagner, Friedrich Pfaff, J. Maedler, John Mueller, J. Hyrtl, Gustav Bischoff, Herman von Meyer, E. von Leonhard, Fr. Aug. Quenstedt, K. E. von Baer, Deluc, L. Hauy, Cuvier, Alex. Brongniart, Binet, Biot, Ampere, Aug. Cauchy, Marcel de Serres, De Blainville, Waterkeyn, Chalmers, Buckland, Whewell, Sedgwick, Fleming, Conybeare, Edward Hitchcock, John Macculloch, Hugh Miller, Benjamin Silliman, and many other great scientists, found no difficulty in being believers in Christ, and profound students of Nature at the same time.[3]

If, now and then, a prominent scientist is an infidel, he was, as a rule, an infidel before he was a scientist; but he did not become an infidel in consequence of true science. Long before Modern Science was talked of,

3. See Bibel und Natur, by Dr. F. Heinrich Reusch, 2d Edition, Freiburg, 1866, pp. 57-61.

infidels existed. About 3,000 years ago the Psalmist said of such: "The fool hath said in his heart, there is no God." Ps. 13, 1. True Science never lead any one to infidelity; but infidels usually endeavor to drag their infidelity into their scientific researches.

Indeed, there exists no real contradiction between any truth of Science and any truth of Divine Revelation; and, for reasons given above, there never can. All so-called contradictions are only imaginary, and are caused either by making unfounded assumptions, or by misinterpreting truths of Nature or of Revelation,—or of both at the same time. *This shows the necessity of examining carefully what exactly are undoubted results—and not mere assumptions of Science, and what exactly are truths of Revelation.*

WHAT CHRISTIANS BELIEVE CONCERNING THE BIBLE.

Since most so-called scientific objections of modern infidels are directed against the Bible, let us consider what we, as Christians, believe concerning the Holy Book.

In the first place,—the Bible is the Inspired Word of God. The late Council of the Vatican remarked on this point: "These books of the Old and New Testament are to be received as sacred and canonical, in their integrity, with all their parts, as they are enumerated in the decree of the said Council (of Trent), and are contained in the ancient Latin edition of the Vulgate. These the Church holds to be sacred and canonical, not because, having been carefully composed by mere human industry, they were afterwards approved by her authority ; nor merely because they contain revelation, with no admixture of error,—but because, *having been written under the inspiration of the Holy Ghost, they have God for their Author,* and have been delivered as such to the Church herself."

The direct consequence of this doctrine, is that nothing false whatever,—not only in matters of faith and morals, but also in all other matters—could have entered into these Holy Books,—whilst they were written under the guidance of the Spirit of Truth.

The next question is : Admitting that the inspired writers wrote down no error whatever—yet, have we, among the many different now existing copies and translations of the Bible, still a version which gives us a reliable text of the Holy Books, as originally written under God's guiding inspiration? This question has been answered by the Council of Trent, which declared that the ancient Latin Vulgate edition of the Bible, which had been tested and approved of by long use, during several centuries, in the Church (quae longo tot saeculorum usu in ipsa ecclesia probata est), is to be considered as authentic (pro authentica habeatur) in public lectures, disputations, sermons, and expositions. Thereby, as the great theologian J. Perrone[4] teaches, the Council declared the intrinsic conformity of the Latin Vulgate translation with the original text not only in matters of faith and morals, but also in other matters, at least substantially (saltem quoad substantiam).

Some of the best Protestant biblical critics, as Bengel, Griesbach, Lachman, and Tischendorf, have, as Rector

---

4. Praelectiones Theologicae, Ratisbonae, 1854, vol. III, p. 125.

M. Heiss, now Archbishop of Milwaukee, observes (in
"The Four Gospels Examined, etc.," 1863, p. 212), after
careful researches come to these conclusions : 1. The
Latin Vulgate is one of the authorities for critics who
endeavor to restore a Greek text, more similar to the
original text than the existing ones. 2. The Vulgate
comes nearer to the real original, than any other modern
translation, according to the so-called original Greek.
Yet, that, in matters of minor importance, owing to
mistakes of copyists and translators, differences exist
between the Vulgate and the original text, which no
more exists, cannot be denied—facts prove it. For
instance, some numbers and proper names have undoubtedly been changed, as Dr. Kaulen, one of our most
prominent biblicists, conclusively shows. He remarks :[5]
" For intrinsic and extrinsic reasons, no change of texts
has been introduced in the Holy Scriptures, which could
disfigure their *essential* contents, especially in matters
of faith and morals. But in the course of investigation
it has been shown, that the form of the text of the Holy
Scriptures is in very many passages not fixed beyond a
doubt. According to Tischendorf's calculation, in the
New Testament alone, more than thirty thousand differences can be found among the various testimonies of
tradition. See his publication: " Haben wir den aech-
ten Schrifttext der Evangelisten und Apostel ?" Leipzig,
1873, S. 12. Though most of these differences may be
unimportant, yet the certainty is thereby taken away
that we have still the text of the Holy Scriptures in the

---

5. Einleitung in die heilige Schrift, Freiburg, 1876, p. 150.

literal form, wherein it was written down by the inspired writers."

If this can be said of the books of the New Testament, we need not be surprised to find, on points of minor importance also, remarkable differences between the text of the Latin Vulgate translation and other existing texts of the Old Testament. For instance, there exists a notable difference between the Septuagint and the Vulgate concerning the times between the creation of man and the Deluge, and between the Deluge and Abraham. Dr. Rohling remarks:[6] "Even the Latin (Roman Catholic) Church has preserved these patent differences in her official books. The Roman Martyrology, for instance, says: In the year 5199, after the creation of the world, in the year 2957, after the Deluge, Jesus Christ was born. The Latins (Roman Catholics) preserve these numbers of the Greek version in their official books, although they accepted the Vulgate translation with the numbers of the still existing Hebrew text (about 4000 years after the creation of man, and 2500 after the Deluge)."

Evidently, the original text, written down under the inspiration of God, cannot be made responsible for these differences, or contradictions; for both the Septuagint and the Vulgate are claimed to be translations of the original text. The blame must consequently rest either *with the copyists or the translators*, who made the mistakes. From all this we infer, that in purely scientific or profane matters, which the Bible mentions only inci-

---

6. Natur und Offenbarung, Organ, etc., Muenster, 1872, p. 101.

dentally, translators and copyists may have made mistakes.

THE BIBLE AND PROFANE SCIENCES.

The next point of interest to us now, is the interpretation of the Holy Scriptures. The Vatican Council declared : *"In matters of faith and morals,* appertaining to the edification of Christian doctrine, that is to be held as the true sense of Holy Scripture which our Holy Mother Church hath held and holds, to whom it belongs to judge of the true sense and interpretation of the Holy Scripture ; and, therefore, that it is permitted to no one to interpret the Sacred Scripture contrary to this sense, nor, likewise, contrary to the unanimous consent of the Fathers."

The meaning of this decision is explained by Dr. J. B. Heinrich[7] as follows : "The authority of the Fathers (the venerable ancient teachers of the Church), as also the authority of the Church, extends only to *matters of faith and morals,* and truths essentially connected with them. Consequently, *purely scientific views* of the Fathers, have no greater value than the scientific principles on which they rest * * * For sufficient reasons, we may reject them (those views), however unanimously they may be held by the Fathers."

Let us, finally, consider the relation in which the Bible stands to truths of the purely profane, or scientific, order, which have no, at least not a direct, bearing on religious truths.

Dr. F. Henry Reusch[8] and Dr. Bernard Schaefer,[9]

7. Dogmatische Theologie, Mainz, 1873, Vol. I, p. 810.
8. Bibel and Natur, pp. 21–35.
9. Bibel und Wissenschaft, Muenster, 1881, p. 4.

claim, with other great theologians, that it is never the object of supernatural Divine Revelation, consequently also not of the Bible, to enrich our purely profane knowledge ;—for the acquirement of which God has given us natural faculties. Therefore, we must assume that the Bible nowhere directly intends to give us instructions on purely scientific subjects.

To this truth another one may be added: In matters of purely profane, or scientific knowledge, the inspired writers, *personally*, were not exempt from the false, or erroneous views of their contemporaries. Some of their views concerning the Earth, the Sun, the stars, etc., may have been not only very defective but even directly false. Moreover, it was not the object of Divine Revelation to correct such false views on matters of purely profane knowledge,—not even as far as the inspired writers themselves were personally concerned. But, on the other hand, *the Spirit of Truth could not sanction with His authority any such errors, on even purely profane matters, in the writings which He inspired.*

We must therefore conclude,—that, although the inspired writers may *personally* have had some erroneous views on matters of purely profane science, yet the Spirit of Truth, under whose guidance they wrote, could not permit them to write down anything false in the Inspired Writings, or the Bible ; for, otherwise the Spirit of Truth would have sanctioned falsehoods ;—which cannot be admitted, being a " *contradictio in terminis.*"

But, it will be urged, the Bible is full of expressions that are inconsistent with our present knowledge of natural phenomena.—In the first place, in such matters of minor importance, translators and copyists may have made mistakes, as explained before. Moreover, in judging passages of the Holy Book, on matters of purely profane knowledge, it must be remembered that these are mentioned only incidentally ; for it was not the object of Divine Inspiration to instruct the inspired writers, or the readers of their writings, on such matters. Therefore, the Spirit of Truth could permit the inspired writers to use some popular phrases on profane matters, then in vogue, which, if strictly and absolutely considered, were incorrect, yet nevertheless relatively, that is in a certain sense,—true.

For instance, the Bible (Gen. 1, 16) calls the Moon one of the "two great lights." Now, strictly speaking, the Moon is no light at all, but only reflects the light of the Sun.—Moreover, the Moon is immensely smaller than countless thousands of fixed stars and planets, which appear to us as quite small. Therefore, strictly considered, it is not true that the Moon is one of the "two great lights" of the universe. Yet, considered relatively, as we see it, it is true ; and it is in this latter sense, the Spirit of Truth permitted the inspired writer to call the Moon one of the "two great lights." Again, we read Josue 10, 13: "The Sun and the Moon stood still." Strictly considered, this, most probably, was not true ; but so it appeared to those present. Therefore, they could say "the Sun and Moon stood still ;" just as we yet say, without telling a falsehood, the Sun rises, moves from east to west, etc.: although we know that it is the Earth that moves around the Sun, and not the Sun that moves around the Earth.

Therefore, whenever we read in the Bible of the Earth as standing firm, of the Sun and stars as moving, etc., we must remember that the Bible does not intend to teach any purely profane or scientific truths, but only incidentally uses popular phrases that are at least relatively, or in a certain sense, true.

It is to be hoped that these explanations will suffice to give the readers correct and clear ideas on the relations of the Bible to matters of profane sciences.

Having considered the doctrines of the Church on these points, let us next consider her practical attitude in relation to purely profane, or scientific, matters.

THE CHURCH AND SCIENCE.

The Church has never by solemn decisions of General Councils, or by definitions *ex cathedra*, interfered in settling *purely* scientific questions. It is not her mission to teach profane sciences, but to teach all things whatsoever Christ has commanded—Matth. 28, 20. Dr. Draper[10] is greatly mistaken, when he asserts of the Church in the fourth century: "The Christian party asserted that *all* knowledge is to be found in the Scriptures and in the traditions of the Church; that, in the written revelation, God had not only given a criterion of truth, but had furnished us *all* that He intended us to know. The Scriptures, therefore, contain the sum, the end of *all* knowledge." Had Dr. Draper, instead of indulging in unfounded assertions, studied the writings of the earlier Christian teachers, he would have found that they did *not* consider the Bible, or

10. History of the Conflict between Religion and Science, p. 52.

the traditions of the Church, as the source of *all* knowledge. They knew full-well that besides the divinely revealed truths, there were yet many other truths of the purely scientific order. For this reason many of the early Christian teachers studied, not only the Bible and the traditions of the Church, but also the writings of profane, even pagan, authors. St. Jerome, for instance, in his letter to the orator Magnus, gives the names of many prominent Christian teachers, from the days of St. Paul to his own time, of whom he says, "that one knows not what to admire more, their profane science or their knowledge of the Sacred Scriptures."

Christian teachers have always maintained that outside of the sphere of the divinely revealed doctrines, there exists yet the vast realm of profane, or purely natural, sciences,—a boundless field in which the human mind may freely exercise itself. The Church has always scrupulously avoided to trespass on this field; she has never interfered with *purely* scientific questions. Whenever the Church took notice of such questions, it was only when these, in some way or another, were intruded upon her doctrine or discipline.

For instance, the Church opposed the idea, now and then mentioned between the fourth and eighth centuries, that there exist antipodes, — as long as it was also assumed that these antipodes were not descendants of Adam, and, consequently, not included in the Fall and Redemption of mankind. For her teaching always was, that in Adam all men have sinned, and Christ suffered for all. We now know that antipodes do exist; yet the Church was right in protesting against the idea that these presumed antipodes were no descendants of Adam. About the purely scientific question whether men could live also on the other side of the Earth, the Church never troubled herself.

The persecution of Galileo is continually brought forward, of course, also by Dr. Draper in his shallow work mentioned before, as an argument to prove that the Church was hostile to Science. It is true that Galileo was to some extent persecuted by authorities of the Church ; as also, that his theory was for a short time condemned. But this was not done by any general Council or by any solemn definition *ex cathedra*, to which alone members of the Church are bound to assent in matters of faith and morals ; but by one of the Sacred Congregations at Rome. It is also true that this Congregation was mistaken in declaring Galileo's theory heretical. Although this declaration seems to have had the sanction of the Pope, yet no Catholic is bound to consider it as a *definition ex cathedra ;* no more than any other of the countless ordinary decisions of the Sacred Congregations, which are regularly given with the sanction of the Pope. Moreover, in as far as the decision on Galileo's theory was a decision on a purely scientific question,—whether it is the Earth or the Sun, that moves— it did not properly come within the sphere of faith and morals, wherein alone the Church, or her visible Head, defining ex cathedra, claims infallibility. The proceedings against Galileo were also more a matter of ecclesiastic discipline, than a question of doctrine. Had not then all Europe been in the midst of the religious agitation caused by the so-called Reformation ; and had not the authorities of the Church feared that Galileo's imprudent and impertinent behaviour would add new troubles to the then existing revolt against the Church, his theory would most probably have remained unnoticed by the Sacred Congregation ; for the same theory, when first published by Copernicus, long before Galileo, was not condemned by the ecclesiastical authorities.

To accuse the Church of hostility to Science, on account of the treatment of Galileo, is just as unfair as to accuse the government of the United States of hostility to the principles of free speech and free press, because during the last war some overzealous politicians and editors were imprisoned for using more liberty of speech and press, than the government then considered safe to allow.

The Church is not, and never was, hostile to Science ; for she is convinced that all true Science has its source in God, and that He cannot teach contradictory truths.

After these introductory explanations, let us proceed to examine those points on which some modern infidels claim to have discovered contradictions between Science and Divine Revelation. After carefully comparing what exactly are the results of Science, and what exactly are Doctrines of Divine Revelation, on those points, all the imaginary contradictions, of which some infidels dream, will, like Macbeth's witches (in Shakespeare), have vanished—

" Into the air : and what seemed corporal, melted
    As breath into the wind."

II. THE ORIGIN OF THE UNIVERSE.

ON beholding the visible universe, the question naturally arises in the mind of every reflecting man : Whence have all these things,—the Sun, the stars, the Earth, the plants, animals, and man,—come ?

Divine Revelation (Genesis, 1, 1,) answers : "In the beginning God created Heaven and Earth ;"—that is, called them—that did not exist before,—into existence.

Some modern infidels claim that they cannot comprehend a creation of something out of nothing. Divine Revelation never taught that God made the world "out of nothing,"—as if "nothing" had been "a raw material," to use one of Mr. Ingersoll's silly phrases. What Divine Revelation really teaches, is—that God once called the universe—which did not exist before,—into existence.

If some infidels cannot comprehend how that was done, we will not blame them for it ; for there are yet numbers of daily observable phenomena which they cannot comprehend. No infidel can, for instance, explain even how a blade of grass grows ; how our will moves our arms, etc. ; how we think ;—what matter, force, life, etc., are. As long, then, as they cannot comprehend such palpable phenomena, they deserve no blame, if their limited intelligence cannot penetrate the mystery of Creation.

A very popular infidel objection is the following, expressed by Mr. Ingersoll in these words : "Nearly all truly scientific minds admit that matter must have existed from eternity. It is indestructible, and the indestructible cannot be created. It is the crowning glory of our century to have demonstrated the indestructibility and eternal persistence of force. Neither matter nor force can be increased nor diminished." Prof. Tyndall also remarks : "As far as the eye of Science has hitherto ranged through Nature, no intrusion of purely creative power into any series of phenomena has ever been observed."

Well, even if "the eye of Science" has not yet discovered any "purely creative power," that is no proof that such a power does not exist. With the most improved microscopes, "the eye of Science" has, for instance, still been unable to detect the mysterious power that makes plants grow and produce flowers and fruit ;—and yet plants continue to do so, without waiting for "the eye of Science" to detect the mysteries of vital force.

Moreover, every Christian may cheerfully admit that, in the existing order of things, neither a particle of matter, nor a minimum of force, in spite of all modern scientific progress, can be annihilated or created by man, or any other created being. Yet, this indestructibility of matter, and this persistance of force, do not prove that matter and force have not been originally created by God; they only show that God has created matter and force so,—that no visible or finite being can either annihilate or create them ;—this the Omnipotent alone can do. The lively imagination of our infidels is running away with their sound common sense, if, from the fact that, in the existing order of things, matter and force cannot be destroyed, they jump to the conclusion that these, therefore, have not been created, but must have existed always.

Some inquisitive infidels are also anxious to know what God did throughout the eternity that preceded the creation of the world.

No doubt, people of small mental calibre would soon feel lonesome, if they would, for a length of time, be without company. Men of vast knowledge and profound reflective talents, on the contrary, never feel less lonesome than when they are alone; as, according to Cicero, P. Scipio Africanus observed : " Numquam minus solus, quam quum solus ; nec minus otiosus, quam quum otiosus sum." If some of our shallow-minded unbelievers soon feel lonesome, when they are alone, the reason can easily be explained ; but they have no right to judge, according to themselves, other men,— men of profound intellectual endowments.

Such men feel not lonesome ; a solitary cell, a silent forest, a quiet starry night, is their most cherished society ; and in such society they gladly miss the society of talkative fellow-beings, not only for hours, but for days, months,—and even for years, as some anachorets have done.

Now, if such men, when alone, feel not lonesome, how infinitely less God, whose eyes are " far brighter than the Sun " (Eccle.: 23, 28); who, with one glance, from all eternity. beheld not only all creatures that, in the course of time, were to spring into existence, but also

His own infinite perfections, which are the source of His most perfect happiness, to which no society of finite creatures could give any substantial addition.

III. ASTRONOMY.

1. THE VISIBLE UNIVERSE.

HAVING answered the infidel objections against the *doctrine* of creation, we will next consider the *works* of creation, and the objections that infidels endeavor to draw from them, against Divine Revelation. Let us commence with the grand visible universe,—the heavens,—above and around us.

At first sight, the Earth appears to be a vast, flat surface, above and around which the Sun, Moon and stars, move daily from east to west. The Sun and Moon appear as "two great lights;"—the stars appear to be of by far lesser size than either the Sun or the Moon. With the help of the telescope and spectroscope, modern Astronomy has given us a better insight into the real nature of the grand visible universe that surrounds us.

We now know that the Earth does not stand still, and that the Sun and the stars do not move around the Earth. We now know that the Earth and several other planets move around the Sun, and that moons move around different planets. The Sun, again, moves around its own axis; and with all the planets of the solar system, and their moons, the Sun, moreover, moves, probably, around some other central point or celestial body, in boundless space.

Little is known of the relations of our solar system to other systems of the sidereal universe. We may therefore consider our solar system as a system for itself, whose center is the Sun. The Sun moves around its own axis, from west to east, in about twenty-five days, as we know from the motion of the Sun-spots.[1] Also the planets (Mercury next to the Sun, then Venus, the Earth, Mars, the small planets, Jupiter, Uranus, and finally Neptune) move in orbits, that increase immensely with their distance from the Sun—from west to east, around the Sun. Again, the moons of the Earth, of Jupiter, Saturn, Uranus, and Neptune, all move from west to east around their respective planets. It is therefore an illusion that the Earth stands still, and that the Sun moves around it; it only seems so, as it would seem to one on a sailing ship, that the coast near by is moving, whereas in reality it is the ship that moves.

Next: what is the distance between the Earth and the Moon, the Sun, and the stars? The Moon revolves around the Earth at a distance of nearly 240,000 miles.[2] The Sun is about 95,000,000 miles distant from the Earth. The different planets, when nearest to the Earth, are between 57,000,000 and 2,761,000,000 miles off. What immense distances these are, may be imagined, when we consider that the diameter of our whole Earth is only a little over 8,000, and its circumference, only about 25,000 miles. But these distances are yet insignificant compared with the distances be-

---

1. Astronomy, by J. Rambosson, translated by C. B. Pitman, New York, p. 74.
2. Astronomy, by Denison Olmstead, revised by E. S. Snell, p. 113.

tween the so-called fixed stars and the Earth. The
French astronomer Camille Flammarion[3] asserts: "As
no star offers a parallax of one second, it follows that
the nearest of the stars is distant from the Earth no less
than 206,265 times 92,000,000 miles. The space which
surrounds the planetary system is void of stars to that
distance at least. The star which is nearest to us, Alpha
of Centaur, has a parallax of 0. "91. Its distance from
Earth is 226,400 times the radius of the Earth's orbit,
or 21,000,000,000,000 miles. This is our *neighbor* star,
and its distance is probably the minimum distance between star and star—21,000,000,000,000 miles." The
same scientist thinks that there are stars whose light
cannot reach us in less than 1,00, 1,000, or 10,000 years,
though light travels at the rate of 185,000 miles per
second.

After these remarks on the distances of stars, we may
form a more correct idea concerning their real size.
Daily observation teaches us that a shrub near by appears taller than a distant oak, and an adjacent hill, larger
than a distant mountain. The farther objects are away,
the smaller they appear to us. The sidereal bodies
appear to us to be small only on account of their immense distance. Let us see what their real sizes are ;—
of a few of them, at least, we have a comparatively
correct knowledge. Some of the planets are immensely
larger than our Earth. For instance, the diameter of
Jupiter is 85,399 miles, which makes it 1,387 times
larger than our globe. Saturn is 746 times and Uranus
73 times larger than our Earth, and the Sun is 354,936

---

3. The Popular Science Monthly, New York; August, 1874. p. 426.

times larger than the seemingly large globe we inhabit.[4] Again, the star Sirius is, according to Flammarion, 2,688 times larger than our Sun. And of such immense suns, there must exist countless thousands among the stars too far distant to give a chance to calculate their distance or size !

Dr. Carl Guettler[5] remarks : "With the naked eye about 5,000, with the telescope over 300,000 stars have been counted. It is calculated that there are in all about 500,000,000,000 of stars ; which calculation, however, is without scientific foundation, on account of the impossibility of penetrating the immense space of the universe."

We know that there must be countless thousands of stars, probably immensely larger than our Sun ; yet exactly how many, no mortal can tell. But, what little we know of the visible universe above us, justifies us in exclaiming : What is our little Earth compared with those numberless, immense globes that traverse the universe ! A leaflet compared with an immense forest,— a grain of sand compared with a mountain range,—a drop of water compared with the ocean !

And hence our modern infidels, instead of admiring the power and greatness of the Creator, suggest the following objections : 1. The Earth, and consequently man, are incomparably insignificant in this immense universe of countless great worlds ; therefore Christians are mistaken in regarding man as the crown, the moral centre, of the visible universe. 2. How can one believe

4. Astronomy, by J. Rambosson, pp. 246, 257. 106.
5. Naturforschung und Bibel, Freiburg, 1877, p. 43.

that God chose this little speck in the vast universe, the
Earth, to assume here human nature?   3. Modern
Astronomy has proved the old Christian views concerning the location of Heaven and hell to have been
erroneous.   Before replying to these objections, let us
examine some other point of interest relating to the
grand visible universe.

2. ARE OTHER VISIBLE WORLDS, BESIDES OUR EARTH, INHABITED BY INTELLIGENT BEINGS SIMILAR TO MAN ?

Far be it from us to deny that God's omnipotence
could, or may, have created intelligent beings to inhabit
the countless celestial bodies which the telescope reveals
to us.   Even the famous astronomer and theologian,
Father Secchi, S. J., in his work on " The Sun," is
inclined to think that the stars are " no uninhabited
deserts," but the abodes of intelligent beings, " capable
of knowing, worshipping, and loving the Creator.   And,
perhaps, these inhabitants of the stars have remained
more faithful than we, in the discharge of their duty
of gratitude towards Him, who has called them from
naught into existence."[1]

Also Prof. Winchell[2] favors this view.   Yet, it is
not our object to deal at present in questions of philosophical speculation, but in results of Modern Science.
And Modern Science has not yet been able to detect any
traces which would indicate that any other world is

1. See Das andere Leben, Von Abbe Elie Merie, Mainz, 1882, p. 200.
2. World-Life or Comparative Geology, Chicago, 1883, pp. 490-508.

inhabited by such living beings as our Earth. What Modern Science can say on the subject is : Among the many thousands of celestial bodies with whose nature the telescope and spectrum-analysis have made us acquainted, we can find hardly any one whereon living beings *similar to those existing on our Earth*, could exist.

Probably, nobody will expect such living beings to exist on the comets—those mysterious strangers from unknown parts of the universe, whose masses are, as Robert S. Ball, Astronomer-Royal of Ireland, says,[3] "almost imponderable." No musquito could find on them enough of tangible substance to stand on,—not to speak of their thermal conditions.

There is also no probability that any living beings exist on the Moon. The best telescopes can find no traces of living beings there. Moreover, no atmosphere surrounds the Moon, and all water that once may have existed on its surface has long since been absorbed by "thirsty-rocks." Prof. Winchell[4] observes : "The total disappearance of water and air from the surface of the Moon may be assumed as evidence of an advanced stage of refrigeration ;" so that such living beings as could get along without air and water would have a fair chance of getting killed by cold.

We next turn our attention to the Sun, in quest of a suitable abode for living beings. But here we meet intense heat. According to Kirchoff, the Sun consists of a solid, or partially liquid, nucleus *in the highest state of incandescence.* Such substances as iron, nickel, cobalt,

3. Popular Science Monthly, June, 1883, p. 245.
4. Geology of the Stars.

magnesium, etc., exist in the Sun only in the state of incandescent vapors. Consequently, any plant, or animal, or man, we know of, would, in less than a second, be burned to invisible atoms on the Sun. The Sun, therefore, seems to be a very uncomfortable place for living in. Now, the author of "Spectrum-Analysis Explained," asserts: "From all the observations thus far made, it may be concluded that at least the brightest stars have a physical constitution similar to that of our Sun. Their light radiates, like that of the Sun, from matter in a state of intense incandescence." Therefore, we must conclude that the fixed stars generally, which are suns like ours, and perhaps most of them immensely larger, are also not fit to be abodes for living beings, such as we are acquainted with.

But were it not possible that some of the planets of our solar system are inhabited by living beings, similar to those found on Earth ?

It would be rash to say that this were impossible ;— for God's omnipotence, no doubt, could create living beings adapted to great extremes of heat and cold. On our own Earth some minute organisms are found that withstand enormous extremes of cold or heat. The experiments of the English scientists W. H. Dallinger and Dr. Drysdale, have shown that the germs of some microscopic animalculae. withstand the enormous heat of 390 degrees F. (water boiling at 212 degrees). It would therefore be very rash to assert that it were impossible that any living beings, similar to those on Earth, do live, or could live, on any of the planets. According to the now, among scientists, generally accepted theory of Laplace, every planet, since it must gradually pass from the incandescent state of the Sun to the frigid present state of the Moon, must attain, at some time, a state similar to the present one of the Earth. With some of the planets, that state may be past; with others, it may still come in the distant future.

Yet the question arises, whether the solar light and heat that reach remoter planets, are sufficient to sustain there such life as flourishes on our globe. Mars receives only $\frac{1}{9}$ as much light and warmth as the Earth; Jupiter, $\frac{1}{25}$; Saturn, $\frac{1}{91}$; Uranus, $\frac{1}{400}$; and Neptune, $\frac{1}{1000}$. But, as some have suggested, a greater density of the atmosphere on these planets might more effectually arrest the radiation of heat. Moreover, as Prof. Winchell remarks,[6] "the solar light on the remoter planets is supplemented by numerous moons, and must, at least, be equal to that of the deep waters and dusky terrestrial situations to which numerous forms of life are found especially adapted."

Thus it seems possible that at least lower forms of vegetable and animal life, could, at some time, exist on those planets; but could the more perfect vegetable and animal types found on Earth, exist there?

This is to be doubted, as far as the largest and most distant planets, Neptune, Uranus, Saturn, and Jupiter, are concerned;—they are too distant from the Sun to ever get a climate somewhat similar to that of our present Earth; and without a similar climate, similar more highly organized types of plants and animals, or of man, could probably not exist. Also the planets nearest

---

6. The Geology of the Stars.

to the Sun, Mercury and Venus, seem to be unfit abodes
for more perfectly organized living beings similar to
those found on Earth.   The intense heat these planets
receive from the Sun, would probably kill all the more
perfect plants and animals, we know of ; and by the time
the Sun will have cooled off sufficiently, so as to send
no more light or heat on Mercury or Venus, than at
present on our Earth, these two planets will be in a state
of perfect refrigeration—like our Moon at present ; consequently, without a drop of water on their surface ; and,
without that, vegetable and animal life, as known to us,
cannot exist.—From all we know of these planets, we
may say that, probably, they will never have a climate
similar to that of the Earth, and, consequently, also no
highly organized plants and animals similar to those
now living on our globe.

The planet most similar to our Earth is Mars ; and if
on any planet, it would be there that we ought to look
for a place fit for living beings similar to those on our
Earth.   The French scientist Flammarion[7] remarks of
this planet, Mars :   1. The polar regions of Mars are
alternately covered with snow, according to the seasons.
2. Clouds and atmospheric currents exist there, as upon
our Earth.—Flammarion even goes so far as to believe
that the continents of Mars are covered with a reddish
vegetation.   But this belief rests only on imagination.
The Civilta Cattolica, which published a series of excel-
lent scientific articles in 1881, remarks on the red color
of the continents of Mars :   "The reason why the continents of Mars appear red is entirely uncertain   *   *   *

7. Astronomy, by J. Rambosson, p. 243.

Most astronomers who treat on this point, think that the continents of Mars appear to us to be of a reddish hue, because the sand or barren soil of the planet is of this color."[8]

Although, then, Mars is among the planets, the one most similar to the Earth in some respects, yet many circumstances exist there quite different from those of our Earth. In the first place, Mars being considerably more distant from the Sun than the Earth, it receives but $\frac{4}{9}$, not half, the light and heat the latter receives. Secondly, Mars having a mass but a tenth that of our globe,[9] its power of attraction, or gravitation, must be proportionally less. Even if Mars had an atmosphere equal to our atmosphere in altitude,—what is not probable,—this atmosphere would be so thin that likely none of the higher plants or animals which thrive on the Earth, could live there. We may add, as Mr. John Pratt[10] observes : "The same set of conditions, in ex-aggerated degree, exist in the minor superior planets, Ceres, Pallas, Juno, etc., while the asteroids are as much out of the question as the comets and meteors. In re-gard to the Jovian and Saturnian satellites, only proba-ble conjecture can be indulged * * * But all the obstacles flowing out of deficient gravitation predicated of Mars, exist in equal degree in these satellites, the largest of which is inferior to Mars in dimensions."

Thus, all things considered, we may say that we know of no planet within our solar system, whose climatic and

8. La Civilta Cattolica, I. Luglio, 1881, pp. 31-32.
9. Astronomy, by J. Rambosson, p. 244.
10. The Popular Science Monthly, June, 1883, p. 206.

other physical conditions can be nearly similar to those of the Earth,—and consequently we may conclude that these planets can, probably, never be inhabited by living beings similar to those on our Earth now.

But could not such a planet—quite similar to the Earth in all respects, exist outside of our solar system ?

God no doubt *could* have created one, or numbers of them ; but we have no right to presume that He *did ;* —and Modern Science has no means to decide this point.

Prof. Winchell[11] remarks : "The majority of the fixed stars, we may fairly conclude, are really other suns ; and, being such, it is almost certain that many of them are encircled by planets in all stages of development, from the self-luminous to the wholly refrigerated."

Even admitting this, we have yet no right to assert that "thousands of these planets must at this moment exist in conditions analogous to those of our Earth," as long as we cannot find any such planets even within our own solar system.

Much less have we any right to assert, as some scientists do, that those thousands of planets *are* "the abodes of organic creatures and thinking intelligences." Our own Earth might have safely arrived at its present stage of geological development, without as much as a leaflet or a worm existing on it ;—and it would have remained uninhabited, had not, *at God's command,* the plants and living creatures, man included, appeared.— But Modern Science has no proof that God has enriched

---

11. Geology of the Stars, Boston.

any other globe, besides our Earth, with plants and animals, or made it the abode of organic and intelligent creatures, like man.

We may therefore conclude, with Mr. John Pratt: "The insignificant little globe called the Earth furnishes the only assurance of the higher forms of life * * * The Earth is not the millionth part of the known matter of our system, and, compared with the space occupied by that system, is far more insignificant than the smallest fleck of foam in the ocean. This tiny island in space does indeed teem with life."—That is about all Modern Science can tell on the subject :—The Earth, indeed, does teem with life ;—but whether there exists, in the vast universe, any other globe that teems with life, similar to that on our Earth, Modern Science does not know. All, then, it can say on the subject, is : 1. Life, similar to that existing on our Earth, is impossible on the comets, on the Sun, and on the countless numbers of so-called fixed stars. 2. It is also, as far as the more perfect plants and animals are concerned, likely impossible on any other planet, or moon, of our solar system. 3. We have no known right for assuming that outside of our solar system there exists any celestial body with climatic and other physical conditions similar to those of our Earth. 4. And, even if such a globe, or globes, should exist, we have not the slightest assurance that God has created any living beings on them.

But here our utilitarian friends, who view all things from the standpoint of profit,—who murmur at the ocean for taking up so much of the surface of the Earth, that might be used with advantage for raising corn to fatten pigs,—or, who grumble at the majestic Alps and other mountains for spoiling many a fine field that could be profitably cultivated for raising potatoes,—will exclaim:—But why did God create those numberless worlds—if there is nobody to live therein?—Like Judas they exclaim: "To what purpose is this waste?"

We reply: "My thoughts are not your thoughts: nor your ways, my ways, saith the Lord." Isaias 55, 8. In creating the world, God had no pecuniary profit in view; his object was infinitely more sublime. When God became man, the angels sang: "Glory to God in the highest: and on Earth peace to men of good will." Luke 2, 14.—So we also say, when God created the universe, His object was, to show His glory outwardly, and to make the creatures, especially the rational ones, happy by partaking of the gifts of His goodness. We read in the Psalms 18, 2: "The heavens show forth the glory of God, and the firmament declared the work of His hands." To reflecting minds, admiring the beauty of the things created, the Holy Scriptures give the advice: "Let them know how much the Lord of them is more beautiful than they; for the first Author of beauty made all those things." Wisdom, 13, 3. And St. Paul, Romans 1, 20, says of God: "The invisible things of Him, from the creation of the world, are clearly seen, being understood by the things that are made; His eternal power also and divinity."

The visible creation is a mirror in which God reflects to intelligent beings His infinite perfection, His power, His goodness, His wisdom, His justice, etc. Man is created for God; and the cause of the happiness and joy man feels in contemplating the visible creation, is, because it reflects, to some extent at least, the infinite beauty and perfections of the Creator. This is the true point of view from which the visible creation ought to be studied. Viewed in this light, we may understand the words of God addressed to Job: "Where wast thou * * * when the morning stars praised me together, and all the sons of God (the angels) made a joyful melody?" Job 38, 4–7.

Whatever short-sighted utilitarians may think of the world,—all creatures continue to fulfill their object,— praise together their Lord and Creator,—and, all the sons of God, angels and God-loving men,—notice a joyful melody rising from their minds, when they contemplate the works of the Creator.

3. MAN, THE CROWN OF THE VISIBLE UNIVERSE, AND THE EARTH, THE SCENE OF THE INCARNATION OF THE SON OF GOD.

We may now proceed to consider certain objections, often advanced by infidels, as to the relation of man to the visible universe, and as to the Earth in relation to the mystery of the Incarnation of the Son of God.

John Wm. Draper[1] remarks: "Seen from the Sun, the Earth dwindles away to a mere speck, a mere dust-mote glittering in his beams * * * Of what consequence, then, can such an almost imperceptible particle be? One might think that it could be removed or even

---

1. History of the Conflict between Religion and Science, New York, 1875, pp. 174–5.

annihilated, and yet never be missed. Of what consequence is one of those human monads, of whom more than a thousand millions swarm on the surface of this all but invisible speck, and of a million of whom scarcely one will leave a trace that he has ever existed? Of what consequence is man, his pleasures or his pains?"

Again, some infidels exclaim: "How can one believe that, out of so many millions of enormous worlds, God chose this insignificant speck in the universe, to assume here human nature?"

Our modern infidels seem to imagine that bulk or size, not quality, is the standard of excellence. Yet, they should not forget that even our prosaic utilitarians consider a handful of gold worth more than many a mountain of common useless rock,—and a jewel, as more precious than thousands of acres of many a marsh-land. Not bulk, but quality, is their standard of value.—Why, then, should they blame God, if He also looks more to quality than to mere bulk or size?

The Sun is, indeed, one of the grandest of visible creatures. But what is this burning fire-ball, perfectly unconscious of God or its own existence, compared with a spiritual, immortal soul that knows and loves God?—Scientists claim that enormous eruptions of incandescent vapours, sometimes reaching "an altitude ten times greater than the diameter of the Earth, or, in other words, of 79,000 miles,"[2] take place on the Sun. A grand and sublime spectacle! some may say.

2. Astronomy by J. Rambosson, p. 83.

But before God, the simple-hearted prayer of an innocent child is more grand, more sublime—and more precious!

·The venerable Thomas A. Kempis[3] remarks: "Indeed, an humble husbandman, that serveth God, is better than a proud philosopher, who neglecting himself, considers the course of the heavens." And we may add: An innocent child that knows, loves, and serves God, is better, and incomparably more precious in the sight of God, than millions of burning fire-balls like our Sun, that are unconscious of their own existence and of the existence of their Creator.

Whatever infidels may say to the contrary, for all Modern Science knows, man may be the *moral centre* of the visible universe in the sight of God.—Although man is not the exact physical point around which the numberless flaming globes and their satellites revolve, yet he is a microcosmos, or miniature world, in whom the material and spiritual worlds unite and centre.

To the objection: "How is it possible to believe that God assumed human nature on this little Earth?"—we answer with Dr. Lorinser:[4] "The incomparably sublime and adorable idea which was realized and expressed by the Incarnation of God, is but that of an infinite self-humiliation of the Divine Majesty, to compensate by this infinite humility for the greatness of pride exhibited by the rebellion of the creature against his infinitely great Creator. Therefore, the main object of the mystery of the Incarnation of God, as of the work of Redemption in general, is based on humility. Therefore

---

3. The Following of Christ, Book I, Chap. ii, 1.
4. Astronomie, etc., Regensburg, 1876, pp. 288-9.

the cradle of the Redeemer was a manger, the house of
His birth a stable, the city of His birth the little Bethlehem, the least among the cities of Juda. And therefore, we may add, the planet upon which this mystery
of humility was to be accomplished, was the Earth, this
little Bethlehem among the sidereal worlds; this manger
of the universe."

Infidels may sneer at this idea; Christians will console themselves with the words of the Apostle St. Paul
to the Corinthians: "The sensual (or animal) man
perceiveth not the things that are of the spirit of God:
for it is foolishness to him, and he cannot understand."
1 Cor., 1, 14. And again: "The wisdom of this world
is foolishness with God." Ibid. 3, 19. And again:
"We preach Christ crucified, to the Jews a stumbling-
block, and to the Gentiles foolishness. But to them
that are called, Christ is the power of God, and the wisdom of God." Ibid. 1, 23, 24.

4. THE LOCATION OF HEAVEN AND HELL.

Infidels often assert that Modern Science has refuted
the Christian idea of Heaven and hell. For instance,
E. DuBois Raymond says of Copernicus, that "he proved
the non-existence of the so-called empyrean, the supposed abode of the heavenly hosts."[1] Dr. Draper
asserts the idea that the Earth is the centre of the
universe, and man the central object of the Earth,—is
"the philosophical basis of various revelations,"—
among which he, of course, intends to include especially
the Christian. Then he adds: "These revelations,

---

1. The Popular Science Monthly, June, 1883, p. 249.

moreover, declare to him (to man) that above the crystalline dome of the sky is a region of eternal light and happiness—Heaven—the abode of God and the angelic hosts, perhaps also, his own abode after death; and beneath the Earth, a region of eternal darkness and misery, the habitation of those that are evil."[2] According to Dr. Draper, Modern Science has, of course, refuted also this Christian idea, — one of the most essential.

Let us see what, according to the opinions of renowned theologians, the Church really teaches about the location of Heaven and hell.

Dr. J. II. Oswald, one of the most profound and reliable theologians of Germany, observes :[3] "Christian terminology, like the Holy Scriptures, calls the abode of the blessed—Heaven; yet thereby it is not yet decided that this signifies a certain location in space; * * * for it is hardly probable, much less proved, that the souls of men in their purely spiritual existence, or even in their glorified bodies after the resurrection, will be confined to a certain location in space * * * Heaven is, in the first place, to be considered only as a *state* of the blessed."

Joannes Perrone, one of the most famous of modern Catholic theologians, says of the location of hell :[4] "This alone is *de fide* (that is, must be believed by all Catholics), that there is a hell, or punishments destined for the wicked, and that these will be without end. All

2. History of the Conflict between Religion and Science, p. 153.
3. Eschatologie, Paderborn, 1869, pp. 38-9.
4. Praelectiones Theologicae, Ratisbonae, 1854, vol. v, p. 253.

the rest, as to the location of hell, or the nature of these punishments, is not *de fide*."—For there exists no final decision of the Church on these points. Long before so-called Modern Science was heard of, St. Augustin remarked[5] on hell: "What kind of fire it is, or in what part of the world or universe it will be, I think no man knows."

From all this we see that the Church has not yet given any final decision as to the location of Heaven and hell; consequently, Modern Science could not refute any really Christian doctrine on this point. The Church simply teaches that there is a Heaven for the good— and a hell for the wicked, where all receive their just reward or punishment; but where these are, is, for the present, known to God and those immediately concerned. In the boundless universe there is, evidently, place enough for both Heaven and hell,—and if Modern Science cannot find out exactly where these are, it is nothing surprising; for there are numberless things yet nearer home, which Modern Science is unable to fathom. For instance, what makes grass grow ?—Or, what makes a worm feel ? etc.

As long as Modern Science is perfectly unable to decide exactly where the centre of the visible universe is, or where its limits are, modern infidels need not be surprised if they are unable to find out the exact location of Heaven or hell ;—about which Divine Revelation did not deem it necessary to gratify the curiosity of some people, who would trouble themselves more about the location of Heaven, than about the means of getting there.

5. De Civitate Dei, Liber 20, Cap. 16.

5. UNITY OF THE VISIBLE UNIVERSE.

In a preceding article we have shown that there exist numberless suns, many of them immensely larger than our own, and many planets much larger than our Earth ; moreover, that these different suns, or, as we often call them, fixed stars, are immensely distant from our Earth and from one another ; so that even light, which travels at a rate of nearly 200,000 miles per second, cannot pass within several—perhaps thousands of years—from many a star to another, or to our Earth.

It may be asked : Does there exist some common bond between these distant worlds ? Is the universe *one grand whole*—or a conglomerate of worlds, without any common ties ?

Modern Science has proved the words of the English poet Alexander Pope, to be true :

"All are but parts of one stupendous whole."

All parts of the visible universe are, we may say, united with one another by an invisible chain,—the force of mutual attraction, or the law of gravitation. This law which pervades the whole material universe, wherever scientific researches can penetrate to, is commonly explained as follows : "Every particle of matter attracts every other particle with a force which varies directly as the product of the masses and inversely as the square of their distances." Upon this law the celebrated system of Newton is based, of which J. Rambosson[1] says : "The summary of his system is this:

---

1. Astronomy, p. 31.

Just as all weighty bodies gravitate to the Earth's centre, so do the bodies which compose the universe (our solar system) gravitate, by the force of attraction, towards the Sun, which is their common centre. But as the planets, if they were only governed by the force of attraction—that is to say, by the force which the Sun exercised in attracting them towards itself, they would gradually be drawn into that celestial body, Newton adduced two moving powers given them by the Creator at the beginning of the world; the first, which was a centripetal force, impelling the planets towards the Sun ; the second, a centrifugal force, which hurried them away from it, the one counterbalancing the other. Thus the Earth, instead of being carried far away from the Sun by the centrifugal force, is maintained by the action of the two combined, in its orbit, and compelled to describe around it an ellipsis of which it occupies one of the foci."

Thus we see how Divine Wisdom has united motion with stability in the universe, and how true the words of Holy Scripture are : "Thou hast ordered all things in measure, and number, and weight." Wisdom, 11, 21.

To illustrate this, let us take, for instance, the motion of our Earth. It is attracted by the Sun with a *certain* amount of force. If this force alone would operate, the Earth would quickly plunge into the Sun. Now, this *certain* force which attracts to the Sun, must be counterbalanced by *exactly* the same amount of force,— to keep the Earth from falling into the Sun. This other force is called the centrifugal force, and is exactly calculated by Divine Wisdom, according to the weight of the Earth, its distance from the Sun, and the velocity of its rotation around the Sun. Were any one of these three data materially changed, the Earth would either fly away from the Sun, or else plunge into it. Were the Earth weightier, or its motion quicker, or its distance from the Sun greater, its centrifugal force would be greater than at present, and it would gradually, in ever enlarging orbits, fly away from the Sun into boundless space. On the contrary, the Earth would, in ever decreasing orbits, finally plunge into the Sun, if either its weight were less, its motion slower, or its distance from the Sun shorter than it is.

The same is true of all the planets of our solar system, and of all moons in regard to their respective planets.

Here we see how the heavens show us numberless cases of Divine Power and Wisdom. Only the fool can say in his heart, there is no God, as the Psalmist observes, if he considers how complicated the motions of the celestial bodies are, and how exactly their relative distances, their masses, and their motions are calculated. No fair-minded astronomer can be an infidel. If some astronomers are infidels, it is only because they brought their prejudiced infidelity into the study of astronomy; but they are not infidels *in consequence* of their knowledge of astronomy.

J. Rombosson[2] remarks: "It is more than a century since Lalande had the hardihood to exclaim: 'I have examined the whole expanse of the heavens, and I can find no trace of God.' And this because he had observed in every direction the traces of infinite wisdom!

2. Astronomy, p. 376.

If the universe had been so imperfect that the Almighty was incessantly occupied in replacing the stars in their courses, His existence would not be called into question; but because His work bears the impress of infinite wisdom, we adduce that very fact as a reason for contesting the existence of the maker! Is it possible to find a more striking instance of unreason? Should we not rather feel that the farther the study of the universe is carried, the more convincing are the proofs of the grandeur and the perfection, not only of it, but of Him who created it?"

To return to our subject: That the Newtonian law of mutual attraction, or gravitation, exists within our whole solar system, no prominent astronomer denies. But does this law also govern the sidereal worlds immensely beyond our solar system?

We reply: The greatest base-line which astronomers have, to calculate the distances of remote stars and their motions, is the orbit of our Earth. Though this seems to be an immense base-line, yet it is so insignificant as to permit the distance of only a few of the more remote stars to be measured; of their motion we know little, or nothing.

Robert S. Ball, Astronomer-Royal of Ireland,[3] observes: "Except for what the binary stars tell us, we would know nothing as to the existence or the non-existence of the law of gravitation beyond the confines of the solar system * * * If we know so little about the existence of gravitation in the space accessible to

---

3. The Popular Science Monthly, May, 1883, pp. 38-39.

our telescopes, what are we to say of those distant regions of space to which our views can never penetrate?"

But, although we have no direct proof, from actual observation, we have yet a very strong circumstantial evidence that the law of mutual attraction, or gravitation, exists throughout the whole visible universe. This evidence is based on the following facts: In the first place, this law exists throughout our whole solar system; in the second place, the sidereal bodies outside of our solar system, consist of essentially the same elementary constituents as our Earth, Sun, etc. This we know from the Spectrum-Analysis.

A prominent scientist[4] remarks: "In the largest instruments (telescopes) the stars remained diskless, never appearing more than as brilliant points * * * Of the *peculiar nature* of these points of light, and of what substances they are composed, the Spectrum-Analysis alone can disclose to us this much coveted knowledge."

Rombosson[5] says on the same subject: "The result of the spectrum studies goes to prove that the stars only vary from each other, and from the Sun, in special and minor ways, and that there are no important and essential differences in their construction. M. Faye, in one of his reports to the French Academy of Sciences, says: 'Thus we see extended to all the stars of the universe that *unity of composition*, which distinguishes our solar world and the aerolites.'"

---

4. Spectrum Analysis Explained, Boston, 1872, pp. 131-3.

5. Astronomy, pp. 64-5.

From this we see how true the words of the English poet are of all the visible worlds, no matter how distant:

"All are but parts of one stupendous whole."

All are composed of essentially the same elements, most of which are also found on our Earth, Sun, etc.; as, hydrogen, sodium, magnesium, iron, etc.;—and are, probably, united with each other by the mysterious bonds of mutual attraction.

As, in the immensity of the visible universe, we may study the omnipresence of God, its Creator and Ruler, so may we also find in the unity of the visible universe, a proof of the Christian Doctrine expressed by the words of the Catholic Creed: "I believe in *one* God * * * the Maker of Heaven and Earth."

How true, then, are the beautiful words of the Psalmist: "The heavens show forth the glory of God."— 18, 1. If one wishes to reflect on the power, wisdom, omnipresence and unity of God, let him study astronomy with a pure and candid mind. There he will learn to understand the meaning of the words of the Apostle to the Romans: "The invisible things of Him, from the creation of the world, are clearly seen, being understood by the things that are made: His eternal power also and divinity." 1, 20.

6. THE FUTURE OF THE VISIBLE UNIVERSE.

On reflecting on the grand visible universe, the following question may come to our mind: "Will this immense and beautiful world remain forever as it is,—or will it once come to an end?"

He that created the universe, and who every moment "upholdeth all things by the word of His power," Hebrews 1, 3, teaches us: "Heaven and Earth shall pass away; but My word shall not pass away." Matth. 24, 35. The Apostle St. Paul writes to the Corinthians: "The figure of this world passeth away." I, 7, 31. The prince of the Apostles writes on the same subject: "The day of the Lord shall come as a thief, in which the heavens shall pass away with great violence; and the elements shall be dissolved with heat; and the Earth, and the works that are in it, shall be burned up * * * But we look for new heavens and a new Earth, according to His promise, in which justice dwelleth." I. St. Peter, 3, 10, 13. Our Lord and Savior foretold: "The Sun shall be darkened; and the Moon shall not give her light; and the stars shall fall from heaven; and the powers of the heavens shall be moved." Matth. 24, 29. St. John writes in the Apocalypse, chapter 21: "And I saw a new heaven, and a new Earth. For the first heaven and the first Earth was passed away; and the sea is no more. And I John saw the holy city, the new Jerusalem, coming down from God and out of heaven, prepared as a bride adorned for her husband. And I heard a great voice from the throne, saying: Behold the tabernacle of God with men; and He will dwell with them: And they shall be His people: and God Himself with them shall be their God: And God shall wipe away all tears from their eyes: and death shall be no more; nor mourning, nor crying, nor sorrow shall be any more; for the former things are passed away. And He who sat on the throne, said: Behold I make all things new." Concerning the new Jerusalem St. John adds: "And I saw no temple in it. For the Lord God Almighty is the temple thereof, and the Lamb. And the city needeth not Sun nor Moon to shine in it: for the glory of God hath enlightened it; and the Lamb is the lamp thereof. And nations shall walk in the light of it."

From these testimonies of Divine Revelation it appears: First, that this present visible universe will not remain always as it is, but once it will pass away; secondly, that it will not be annihilated, but dissolved by fire; and thirdly, that a new Earth and visible universe will take the place of the present one.

Now let us see what Modern Science has to say on the future of the visible universe. Its verdict is: This visible universe carries the germs of dissolution in itself, and must necessarily once come to end.

In the first place, even if no unforseen catastrophe should occur,—according to certain physical laws,—no matter after how long a time,—the moment would once inevitably arrive, when the now blazing stars and our fiery Sun would become as cold and frozen, as our Moon is at present; just as surely as the moment must once arrive for any other fire, no matter how great, to become extinct. "Hemholtz says:[6] 'The inexorable laws of mechanism show that the store of heat in the Sun must be finally exhausted.'" Of course, long before the Sun would have become entirely cooled, all

---

6. Sketches of Creation, New York, 1871, p. 411.

living beings we are acquainted with, would have been killed by cold on our Earth.

But some fearful catastrophe may overtake our world before the Sun will have time to become cold. Prof. Winchell[7] calls attention to the remarkable retardations observed in the return of the comets of Encke and Faye. Some scientists are of the opinion that these retardations can be explained only on the assumption that there exists some "allpervading resisting medium commonly called ether,"—the same through which also the light of the most distant stars is conveyed to our Earth. Prof. Winchell asserts: "The proof of the existence of a resisting ether in space, has disclosed the decree which records the doom of the solar system." Whewell remarks: "Since there is such a retarding force perpetually acting, however slight it be, *it must in the end destroy all celestial motion.*" * * * Comte says: "In a future too remote to be assigned, all the bodies of our system must be united to the solar mass, from which it is probable they proceeded."

The verdict of Modern Science, then, is, that according to fixed natural laws now operating,—our Earth, as all other planets, must once fall upon the Sun—and probably burn to atoms,—even if by that time the Sun had already become cool. For, as Prof. Winchell remarks: "Arrested motion becomes heat. A meteorite falling through the Earth's atmosphere develops so much friction as to generate heat sufficient to dissipate the

---

7. Ibid., pp. 417-20.

body into vapor."[8]   How much more the Earth falling
upon the Sun !

But does the resisting ether threaten destruction only
to our solar system—and not also to the universe outside of it ?

The same reasons which prove that this resisting ether
must gradually cause the planets to fall upon the Sun,
also prove that it must cause the other celestial bodies,
by retarding their motion, to fall either upon their common centre, around which, as some astronomers assume,
they move,—or upon one another. For, in the first
place, no doubt, all so-called fixed stars have their own
peculiar motion, although we can perceive little of it.[9]
And, secondly, all move in that resisting ether through
which their light comes to us. No matter how long it
may take,—according to fixed physical laws, the balance
will once be destroyed, which kept the stars in their
orbits,—and, owing to retarded motion, the stars must
once rush together towards their common centre, or fall
upon one another.

Thus we see that Modern Science agrees with Divine
Revelation on the following points :   First, this visible
universe will once pass away ; secondly, most probably
an immense conflagration will take place, by the falling
of the planets upon the Sun,—or of the stars upon one
another ; thirdly, according to the generally admitted
law, "that no particle of matter in the present order of
nature is destroyed"—a new world will spring from
the old, — of course, under God's guiding providence.

8. Ibid., p. 410.
9. See the article "Drifting of the Stars," by R. A. Proctor, in
The Popular Science Monthly, December, 1872, pp. 224-32.

7.   THE THEORY OF LAPLACE ON THE EVOLUTION OF THE VISIBLE UNIVERSE.

Having reflected on the distant future of our visible universe, let us next cast a glance at its by-gone ages,—compared with which the ages of mankind dwindle into insignificant moments.

Has God created *at once* the visible universe,—Sun, Moon, and stars,—just as we see them now? Or have they undergone changes, before arriving at their present condition?

Divine Revelation teaches that God has created all things,—but not the manner in which it was done. Hence, Science has here a perfectly free field for inquiry.

Modern astronomers have generally accepted the so-called theory of Laplace, according to which[1] the matter of our solar system has existed originally "at such a temperature, as to be in the condition of vapor of great tenuity." This vapor gradually cooled and contracted, and thereby started a rotation which, in the course of time, caused rings to form, which separated from the main solar mass; and, being broken at some points, they gradually became planets. Other rings separated from the planetary masses, and became the satellites, or moons of the latter.

This, in brief, is the famous theory of Laplace. Let us see what reasons are advanced in its favor.

Herbert Spencer[2] observes: "Organic progress consists in a change from the homogeneous to the hete-

---

1. Geology of the Stars.
2. Progress: Its Law and Cause.

geneous. * * * The series of changes gone through
during the development of a seed into a tree, or an ovum
into an animal, constitute an advance from homogeneity
of structure to heterogeneity of structure. In its primary stage, every germ consists of a substance that is
uniform throughout, both in texture and chemical composition. The first step is the appearance of a difference
between two parts of this substance. * * * This
process (of differentiation) is continually repeated * *
* and by endless such differentiations there is finally
produced that complex combination of tissues and organs
constituting the adult animal or plant. This is the history of all organisms whatever * * * This law of
organic progress is the law of all progress. Whether it
be in the development of the Earth, in the development
of life upon its surface, in the development of society,
of government, of manufactures, of commerce, of language, literature, science, art, this same evolution of
the simple into the complex, through successive differentiations, holds throughout. From the earliest traceable cosmical changes down to the latest results of
civilization, we shall find that the transformation of the
homogeneous into the heterogeneous, is that in which
progress essentially consists."

These words of the famous English thinker, contain
in a nutshell the, at present, very popular doctrine of
evolution; which is highly extolled by some as a welcome theory, to make the Creator superfluous in the
universe,—and decidedly combatted by others, as an
unchristian and atheistic doctrine.

What shall we think of it? No doubt, this doctrine has been greatly misused by infidels; yet, it may be substantially correct—and it is not inconsistent with any Christian doctrine. In fact, centuries before Herbert Spencer, Laplace, or Darwin, saw daylight, the perhaps greatest and most philosophic Father of Church, the famous St. Augustin, plainly taught, in his work "De Genesi ad Literam," the doctrine of evolution. Dr. Carl Guettler[3] says: "The idea of St. Augustin,—that matter was originally (created) formless, but endowed with the capacity of producing out of itself forms, is nothing but the philosophic foundation of the same theory (of evolution) which Kant and Laplace, 1400 years later, applied to Astronomy, and Lamarck and Darwin, to organic nature."

St. Augustin having openly taught the doctrine of evolution, no Christian need scruple to accept it, if he knows of sufficient reasons in its favor. Because some infidels make a bad use of this theory, some overzealous Christians imprudently jump to the other extreme, and treat it as an infidel invention—hostile to the Christian idea of a Creator. This theory is far from detracting from the dignity of God as Creator and Ruler of the universe;—on the contrary, it gives to the works of the Creator a new and sublimely grand aspect. I, for one, heartily agree with Mr. Le Conte, Professor in the University of California, who observes on the theory of Laplace, what may be said of the theory of evolution in general: "Does not the cheering doctrine of final

---

3. Naturforschung und Bibel, 1877, p. 55. See also p. 145.

causes —of design and purpose—become strengthened and invigorated by leading us to a view so comprehensive ? 'How simple the means—how multiform the effects—how far reaching and grand the design !' How deeply they impress us with the wisdom, power, and glory of the Creator and Governor of the universe !"[4]

This much about the theory of evolution in general ; now as to the theory of Laplace,—or the theory of evolution as applied to the sidereal universe,—at least to our solar system.

The Jesuit Father Secchi of the Roman College, both one of the greatest modern astronomers and one of the most learned Catholic priests of our century, observes in his work on the *Sun ;* "Les savants sont de nos jours *unanimes* a admettre que notre systeme solaire est du a la condensation d'une nebuleuse qui etendait autrefois au-dela des limites occupees actuellement par les planetes le plus lointaines." Le Soliel, p. 332[5]. "Scientists are, now-a-days, *unanimous* in admitting that our solar system is due to the condensation of a nebula which once extended beyond the limits at present occupied by the most distant planets."

Since this theory is accepted as well founded as to our solar system, we may presume that it also applies to the most distant stars ; for they also have a constitution quite similar to that of our Sun.

Let us see on what facts this theory is based. In the first place, all the planets revolve around the Sun as their centre,—all move from west to east, and in approx-

---

4. The Popular Science Monthly, April, 1873, pp. 655-6.

5. Quoted by Alex. Winchell, in "Geology of the Stars."

imately the same plane. In a similar manner, the moons revolve around their respective planets. Moreover, the Sun itself revolves around its axis—also from west to east; as its so-called spots show. Secondly, in our solar system we have illustrations of all stages of planetary development,—from the blazing Sun, through the more or less cooled planets, to the entirely cooled Moon. The Earth still retains much of its original heat, as volcanoes, hot-springs, etc., prove; while some of the larger planets, Jupiter and Saturn, are still in a state of intense heat.—Moreover, the rings of Saturn, the various moons, and the orbits of the planets, indicate a regular, *successive* evolution of our solar system from primeval matter. And last, but not least, modern Astronomy has discovered gaseous nebulae,—such as Laplace's theory postulated. These nebulae are so far away, that we have no means of getting an idea of their distance. Perhaps it takes their light many thousands of years to reach our globe; perhaps, they have long ago ceased to be nebulae, and become suns like ours, and we see them yet on Earth, as they were thousands of years ago,—perhaps soon after creation.

According to the opinion of the learned Jesuit Father Secchi,[6] the discovery of these nebulae has confirmed, if not proved, the theory of Laplace. There is no reason why any Christian should be hostile to this theory. Whatever some shortsighted infidels may say, this theory does not exclude, but on the contrary implies the guiding providence of the Creator, Who has,

6. See Geology of the Stars, by Prof. Winchell.

in this simple manner, produced the most stupendous and complicated effects,—and carried out His "far reaching and grand design."

IV. GEOLOGY AND PALEONTOLOGY.

1. THE SIX SO-CALLED DAYS OF CREATION.

HAVING declared that God created both the heavens and the Earth, the Bible relates the gradual perfection and ornamentation of our globe, up to the time of the creation of man. Here the Bible distinguishes six different periods of time, called "Days."

On the first Day, light and darkness, day and night commenced to alternate ; on the second, the vast expanse we call the firmament, which separates the waters of the clouds from the waters on the surface of the Earth, became discernable ; on the third, dry land began to emerge from the waters that covered the Earth ; and, finally, on the fourth Day, the Sun, the Moon, and the stars, became visible on the surface of the Earth.

Moreover, after the Earth was sufficiently prepared to receive organic beings, plants appeared on the third Day ; animals living in water, and birds, on the fifth Day ; animals living on dry land—and finally man, on the sixth Day.

Now, what have our modern unbelievers to object against these six so-called Days ? They say : Long before the first man existed,—thousands of years before,

—the Earth had been filled with countless species of plants and animals, as their abundant remains in ancient rocks testify ; therefore, the Bible is greatly mistaken in teaching that all this could have happened within the short time of six days.

Our modern infidels must be told again and again, that the Bible does not say what kind of Days these six were, or how long they lasted.

Let it be remembered that these Days were not days of man,—who was created at the end of the last of these six Days, but days of God, the Creator,—before whom "a thousand years are as one day," (I. Peter, 3, 8) ; in comparison with whose eternity, the longest periods of time imaginable dwindle into insignificance. The seventh Day, on which, as the Bible, Genesis, 2, 2, relates, God finished the work of creation, and rested, still continues, after thousands of years ; the Bible gives no intimation of its having ended.

Why, then, could not the other six Days also have lasted for thousands of years ; which to mortal man may seem an immense period, but which are to an *eternal* God less than a day to us ? That the Bible, in using in Genesis I. the word "Day," does not teach it was a duration of time of twenty-four hours, may be inferred from Genesis 2, 1-4, where we read : "So the heavens and the Earth were finished, and all the furniture (or ornament—"ornatus," as the Vulgate has it) of them. * * * These are the generations of the heaven and the Earth, when they were *created in the Day* that the Lord God made the heavens and the Earth." Here we see the word "Day" used in the sense of a period of time, including all those "six Days." We are, therefore, justified in assuming that the Bible, in speaking of those "six Days," did not speak of a duration of time of only twenty-four hours, but of indefinite periods of time, that lasted for, what we would call, many ages; — whose "evenings" and "mornings," — beginnings and endings, may have lasted longer than all the historical ages of mankind.

For such reasons, not only the learned ancient Jewish writer Philo, but also ancient Christian writers, as Clement of Alexandria, Origen, and Procopius, considered those six Days not as ordinary days of twenty-four hours, but as periods of time of unknown duration. The Church has never given any decision on this point, and so we are free to say, as the great teacher St. Augustin, who lived about fifteen centuries ago, said: "What kind of Days they are, is either most difficult or impossible for us to guess; how much more so to decide."

Therefore Christians, as far as their religion is concerned, need not object, if modern scientists teach that many thousands of years ago, — long before the first man appeared, — the Earth had been filled with countless species of plants and animals.

2. THE ORIGIN OF THE MOSAIC ACCOUNT OF THE CREATION.

To understand correctly the history of the Creation, as related in the first chapter of the Inspired Book, it is necessary to consider the origin of the so-called Mosaic Account.

That the Book of Genesis is one of the books written under the inspiration of the Spirit of Truth, has always been the doctrine of the Church. Now, as theologians teach, the Spirit of Truth may have impelled inspired writers to write down correctly, either some revealed truths which God alone could know,—or some truths or facts, which the inspired writers themselves had learned. For instance, the Evangelists were impelled by the Holy Spirit to write down concerning Jesus, what they themselves had either seen, or heard of others ; whereas, St. John, in describing in the Apocalypse the future history of the Church on Earth, from the Apostolic times till to her final triumph in Heaven, wrote something which no man could have known ; which, consequently, the Spirit of God must have revealed to him.

Now, does the Mosaic Account of the Creation relate what the inspired writer himself, or any other reliable person, had witnessed ; or does it contain a revelation from God of truths unknown to man ? Evidently the latter. Man having been created last of the visible world, he could not have witnessed the acts of creation, that preceded his own creation. Nor have we any reason for assuming that thousands of years ago mankind had already made such progress in natural sciences, that they could give such an accurate account of the progressive stages of the creation of our Earth, as we find in the Book of Genesis.

To whom did God make this revelation ? Perhaps to Moses, during the forty days when he was on Mount Sinai with the Lord,—or, at some other time, in the

desert ?   Perhaps.—But, most probably, this revelation
was first made to Adam, after his creation. This
view which is held, among others, by the learned Jesuit-
Father F. von Hummelauer, is corroborated by the fact
that, as Dr. Lueken in his "Stiftungs-Urkunde des
Menschengeschlechts" shows, this Account of the Creation was known not only to the Israelites, but, we may
say, to all ancient nations. This can be explained only
on the assumption that this Account reaches back to
the source of all nations—the first parents,—and that it
was preserved by the traditions of the various nations,
until, finally, under the guidance of the Spirit of God,
it was written down in the Book of Genesis by Moses.

The learned Jesuit-Father mentioned before, explains
the origin of the Mosaic Account of the Creation substantially as follows :   Since no man had witnessed the
acts of creation (man having been created last), the
Account of the Creation we find in Genesis, Chapter I.,
necessarily postulates a divine revelation. To whom
was this revelation made ?   The Book of Genesis relates (2, 3) that, on the seventh Day of creation, the law
of keeping the Sabbath holy was instituted. This law
stands in such an intimate connection with the "six
Days of Creation," that we are justified in assuming
that the commandment to keep the Sabbath holy, was
given to Adam at the same time when the history of
creation was revealed to him. The traditions of the
most ancient nations unanimously point to the fact that
this revelation was originally made known to all mankind.   These traditions agree, not only in the leading

ideas, but also in minute details, with the account written in the Book of Genesis (as Dr. Lueken has clearly shown in his work, " Die Stiftungsurkunde des Menschengeschlechts "). The Account of the Creation we have, seems, therefore, to be *the description of a vision which Adam had,* wherein the acts of creation were shown to him by God. Under the representation of Days, the long geological ages were shown to Adam. In ecstacy, as St. Paul, when Paradise was shown to him, I. Cor. 12, or as St. John, when the future history of the Church was revealed to him, Apoc. 1, 10, Adam witnessed the progress of creation. *At the word of God,* he saw light penetrating the darkness surrounding the Earth ; he saw an expanse, called the firmament, forming between the waters on the Earth and the waters in the clouds ; he saw dry land emerge from the waters covering our globe ; he saw plants and trees commencing to grow ; he saw the Sun, Moon, and stars, sending their first rays to the surface of the Earth ; he saw animals of every description moving in the waters and springing from the soil ; and, finally, he saw in this vision the process of his own creation. The horizon of his vision was most probably limited. Likely he did not see the progressive perfection of the whole surface of the Earth, from pole to pole. What was shown to him, was, probably, only a certain country of Asia, the Land of Eden ; the same, where he found himself after awaking from this ecstacy, or vision. The successive acts of God, then, in making this country, the first to be inhabited, a fit abode for man, were shown to Adam in six grand panoramic views.

Now, a very important question is the following: Are we obliged to assume, because to Adam these creative acts were thus shown, that they extended over the whole globe in the same relative succession?

Father von Hummelauer[1] is of the opinion that although the history of the creation of the Earth may have been represented thus in six grand views to Adam, yet it is not necessary to assume that all over the Earth the same series of events succeeded each other in the order mentioned in the first chapter of Genesis. Perhaps, in some places aquatic animals were created on the third day; perhaps, in some places, fowl appeared on the fourth day; perhaps, in some places, even animals liv-ing on land, appeared on the fourth day.

The point of view, then, which we must not lose sight of, is: Adam saw in a vision the successive acts of creation in the order they succeeded each other in the Land of Eden, the first home of mankind; but that these acts of creation extended in the same order *all over the globe*, we are not compelled to assume from the Account of the Creation given by the Bible.

Moreover, since it was not the object of God's Revelation to give mankind a *scientific* instruction on the origin of the Earth and all vegetable and animal species that ever existed, but only to instruct man that God is the Creator of the *whole* visible universe,—we may assume that not the origin of *all* species of plants and animals that ever had existed in the Land of Eden, but the origin of only those species of plants and animals

---

1. Der biblesche Schoepfungsbericht, Freiburg, 1877.

which still existed at the time of Adam, were shown to him in the vision.

This vision of Adam was handed down by tradition among his descendants, till Moses, inspired and guided by the Spirit of God, recorded it for the instruction of all future mankind. Hence it is often called the Mosaic Account of the Creation.

Such, then, is the theory held by some prominent theologians. Of course, it is a "theory," or hypothesis; yet such solid reasons seem to be in its favor, that it may be considered as one of the most plausible theories which have been advanced, to serve as a key for the explanation of the Biblical history of the Creation.

Without claiming this to be the only theory—for there are yet others—on which the Mosaic Account of the Creation and the results of Modern Science can be satisfactorily reconciled, — we shall adopt it for the present in the following explanations.

3. THE MOSAIC ACCOUNT OF THE CREATION, COMPARED WITH THE RESULTS OF MODERN SCIENTIFIC INVESTIGATIONS.

After declaring in a few words, that God in the beginning, created the inorganic elements of the heavens and of the Earth, Moses turns his entire attention to the history of the gradual perfection and ornamentation of our globe. The first condition of the Earth which he describes, was as follows: "The Earth was void and empty, and darkness was upon the face of the deep: and the Spirit of God moved over the waters. And God said: Be light made. And light was made.

And God divided the light from the darkness. And He called the light Day and the darkness Night." Genesis 1, 2–5. This was the

FIRST DAY

of the gradual perfection of the Earth.

With what stage of the evolution of our globe, according to the results of modern scientific investigations, does this "first Day" correspond?

The reader will remember that according to the theory of Laplace, now generally accepted among scientists, our Earth was once in an incandescent state, as the Sun is still. At that time, the Earth was a shining sun, and no difference of day and night could be noticed on its surface. But since the Earth is by far smaller than the Sun, she, in the course of time, lost her heat sooner, and ceased to be self-luminous; whilst the Sun still continues to shine in all his brightness. In proportion as the Earth lost her billiant glare, dense vapors, arising from the intensely hot surface, commenced to surround her. Then, first, the alternation of day and night (of light and darkness) could be noticed. But the dense vapors, surrounding the Earth, made the Sun, Moon and stars, yet for ages, invisible from the surface of our globe. Prof. Winchell[1] remarks of this period: "The Sun rose in the morning, and sent a lurid ray through the dense, refractive atmosphere, and at night sank into the smoke that ascended from a burning world. The morning and evening twilight almost met each other in the

---

1. Sketches of Creation.

midnight zenith, so high and refractive was the heterogeneous atmosphere."

This stage, then, of the Earth's development, seems
to have been the first Day, described in Genesis 1, 3--5.

THE SECOND DAY

is thus described by Moses : "God made a firmament,
and divided the waters that were under the firmament,
from those that were above the firmament." Genesis 1, 7.

It must be remembered that, according to the theory
of Laplace, the Earth, after ceasing to be a self-luminous body, and being surrounded by dense vapors, yet for a
long time retained such a high temperature on her surface, as to change water and other volatile substances
into hot vapors. All the water, carbon, sulphur, etc., on
the Earth, were converted into vapors, which continually
ascended from the heated surface of our globe to the
cooler heights of the atmosphere, to return again in a
condensed shape, and to undergo anew the former
changes, as soon as they came near enough to the hot
surface of our globe.

Thus for a geological age a violent storm, caused by
the intense heat of our planet and the cold existing beyond its atmosphere, continued to rage and to fill all
space between the two thermal extremes with confusion.

Consequently, then the expanse, or what we call the firmament, which at present separates the waters upon the
Earth from the waters in the clouds, did not yet exist.
"The first drop of water," Louis Figuier remarks,[2] "which
fell upon the still heated terrestrial sphere, marked a new

2. The World before the Deluge, New York, 1872.

period in its evolution * * * How long did this struggle for supremacy between fire and water continue ? All that can be said in reply is, that a time came, when water was triumphant. After having covered vast areas on the surface of the Earth, it finally occupied and entirely covered the whole surface * * * The ocean was universal."

The more the cooling of our globe went on, the more, gradually, that expanse, often called the firmament, became discernible, of which the Bible says that on "the second Day" it "divided the waters that were under the firmament, from those that were above the firmament," —that is, the waters on the surface of our globe—from the waters in the clouds.

OF THE THIRD DAY

the Book of Genesis relates : "God said : Let the waters that are under the heaven, be gathered into one place : and let the dry land appear. And it was so done." Genesis 1, 9. — As mentioned before, there was a time when, according to the opinion of geologists, the surface of our Earth was entirely covered with water ; yet, under God's guiding providence, dry land, the future abode of mankind, was, in the course of time, to emerge from the waters. — A crust, covered with water, had been formed on the surface of our globe. But since the refrigeration, or loss of heat, continued as Prof. Winchell observes, the stiffening crust would become too large for the nucleus within. The crust, therefore, must wrinkle, to fit the shrinking nucleus. Thus, incipient inequalities of the surface began to appear. — These were the germs of mountains and of continents, which in the course of ages gradually became larger. Thus, as the Book of Genesis relates, the waters gathered together (into oceans, etc.), and dry land began to appear.

ON THE FOURTH DAY,

according to the Hebrew text and the Vulgate translation, God said : "Let there be lights (fiant luminaria) in the firmament of heaven, to divide the day and the night, and let them be for signs, and for seasons, and for days and years. And God made two great lights : a greater light to rule the day, and a lesser light to rule the night : and the stars." Genesis 1, 14, 16.

Let it be remembered, the Bible teaches that God, in the beginning, had created the heavens and the Earth,— that is the *elements* of our Earth — and of the Sun, Moon, and stars;—but these *became visible* on the surface of our present globe only on and since the fourth Day. — We may therefore assume that long before the fourth Day these celestial bodies had existed, perhaps, about in the same perfection as at present, yet, before the "fourth Day" they could not be seen on the surface of our Earth, since she was still surrounded by dense vapors and clouds, which prevented the rays of sidereal bodies from penetrating to her surface.

The Earth continued to lose her surface-heat ; the rising vapors grew less ; and the dense, dark clouds gradually disappeared and permitted the light of the outside worlds to reach the very surface of our globe. Then the normal relations between our Earth and the Sun, Moon, and the stars, commenced to exist, as described, Gen. 1, 14, 16.

From this we see how admirably the results of Modern Science agree with what the Bible teaches on the gradual development, or perfection of our globe, and its relations to other sidereal bodies.

But our Earth is not only a globe consisting of various inorganic, elementary masses, but also the abode and substratum of organic beings, plants and animals, —with their visible crown—man. What, then, have Modern Science and the Bible to say on the appearance of plants, animals, and man, on our globe?

The Book of Genesis 1, 11–27, teaches that, after dry land had appeared on the third Day, God said: " Let the Earth bring forth the green herb, and such as may seed, and the fruit-tree yielding fruit after its kind." On the fifth Day, after the Sun, Moon, and stars, had become visible in all their brightness upon the surface of the Earth, God said: " Let the waters bring forth the creeping creature having life, and the fowl." On the sixth Day, God said : " Let the Earth bring forth the living creature in its kind, cattle, and creeping things, and beasts of the Earth according to their kinds." Finally, God said : " Let us make man to our image and likeness : and let him have dominion over the fishes of the sea, and the fowls of the air, and the beasts, and the whole Earth, and every creeping creature that moveth upon the Earth." According to the account of the Bible, then, the order in which organic and living beings appeared on the Earth, was as follows: Plants appeared first,— on the third Day; next appeared animals living in water, and fowl,—on the fifth Day; then appeared animals living on land, and, finally, man,—on the sixth Day.

Let us see what Modern Science has to say on this subject.—Mr. Archibald Geikie[3] calls attention to the volcanoes, the hot springs, and the increase of temperature towards the centre of the Earth, which seem to indicate—"that the inside of our planet must be in an intensely heated condition;" in fact, if, as observation seems to establish, "the temperature rises about 1 degree Fahrenheit for every fifty or sixty feet of descent,"—then it is quite probable that "at the depth of about two miles water would be at its boiling-point, and at depths of twenty-five or thirty miles, the metals would have the same temperatures as those at which they respectively melt on the surface of the Earth." The same author continues: "There can be no doubt that, at one time, many millions of years ago, the globe was immensely hotter than it is now. In fact, it then resembled our burning sun, of which it once probably formed a part, and from which it and the other planets were one by one detached. During the vast interval which has passed away since then, it has been gradually cooling, and thus the heat on the inside is only the remains of that fierce heat which once marked the whole planet. The outer parts have cooled and become solid, but they are bad conductors of heat, and allow the heat from the inside to pass away into space only with extreme slowness."

---

3. Geology, New York, 1883, pp. 84–89.

Keeping these facts in mind, one will easily perceive that there must have been a time once, — and a long time — when nothing living, plant or animal, could have existed on our Earth : the intense heat would have melted every organic being to atoms. — Before organic beings, plants or animals, such as we are acquainted with, could exist on our globe, it was necessary that its surface should have cooled considerably. At what time exactly, the first microscopic plants or animalculae could appear on Earth, we know not; all what scientific researches teach on this point, is, that in those rocks which were once in a molten state, we would look in vain for vestiges of either plants or animals ; such are found only in rocks of later formations—in rocks that were formed by the sediments of water. Of course, the relatively lower the strata of these sedimentary rocks are, the more ancient they are, as also the remains of plants and animals imbedded therein.

In what order, then, as far as we can learn from remains imbedded in these rocks, did organic beings appear on our globe? Louis Figuier, in the work mentioned before, remarks : "Did plants precede animals ? We know not; but such would appear to have been the order of creation. It is certain that in the sediments of the oldest seas and in the vestiges which remain to us of the earliest ages of organic life on the globe, we find both plants and animals of advanced organizations. But, on the other hand, during the greater part of the primary epoch—especially during the carboniferous age —the plants are particularly numerous, and terrestrial animals scarcely show themselves; this would lead us to the conclusion that plants preceded animals. It may be remarked, besides, that from their cellular nature, and their looser tissues composed of elements readily affected by the air, the first plants could be easily destroyed without leaving any material vestiges; from which it may be concluded, that, in those primitive times, an immense number of plants existed, no traces of which now remain to us." Professor Winchell in his "Sketches of Creation" declares it "unsafe to determine at which epoch the waters of the primeval sea became sufficiently cooled and purified to receive the first organic forms." * * * But, "reasoning deductively, it is presumable that vegetable life preceded animal life in order of appearance."—For "vegetable life is capable of enduring more extreme conditions. *Vegetation, moreover, is capable of drawing its sustenance from the mineral world, while animals rely exclusively upon organic food.*"

Therefore, the scientist just mentioned concludes : "All things considered, we are led to believe that plant life had a history upon our Earth, a full epoch before the existence of the lowest animal."

Thus we see, how the results of modern scientific investigations establish what Moses, inspired by God, taught thousands of years ago, namely, that plants were created before animals.

But it is to be mentioned that in the lowest (oldest) strata, wherein remains of organized beings are found, as also in the succeeding and later strata, up to the time of man, remains of plants and animals are *found together ;* not plants alone first, and animals later, as one would, at first sight, be inclined to expect from the account given by Moses, who relates that God created the plants on the third,—and the animals on the fifth and sixth Days.

"It is now well known," Mr. Ingersoll observes, [4] "that the organic history of the Earth can properly be divided into five epochs—the Primordeal, Primary, Secondary, Tertiary and Quaternary. Each of these epochs is characterized by animal and vegetable life, peculiar to itself. It is reasonable to suppose that certain kinds of vegetation, and certain kinds of animals should exist together, and that, as the character of the vegetation changed, a corresponding change would take place in the animal world."

How can this seeming contradiction between the Mosaic account, which describes the plants as having been created on the third Day—and the animals, on the fifth and sixth Days — be reconciled with the results of paleontological researches, according to which, as also Father von Hummelauer observes, [5] "every new geological epoch had a flora and fauna (species of plants and animals) corresponding with the existing climatic conditions"?

We must remember, Divine Revelation did not intend to give man a scientific instruction on the precise order in which plants and animals appeared all over the globe; its main object was to teach man that God

4. Some Mistakes of Moses, Washington, D. C., 1882, pp. 69, 70.
5. Der Biblische Schoepfungsbericht,

is the Creator of *all* things visible; and this object was completely attained by revealing the different acts of creation,—in six grand panoramic views, which had probably been shown to Adam in a vision. Dr. Guettler[6] justly observes: "Whether the Sun appeared in its full brightness before or after the plants; whether the animals were created after the plants, or not; how long those "Days" lasted; how they coexisted or succeeded one another,—all these are questions of minor importance to the Bible (or Divine Revelation). The main object of the Bible is to represent to us clearly, in different acts, God as the Creator of the whole world."

Some unbelievers ridicule Moses for relating the creation of the plants before the Sun was visible on the Earth. Mr. Ingersoll[7] remarks: "It does not seem to me that grass and trees could grow and ripen into seed and fruit without the Sun. * * * After the world was covered with vegetation, it occurred to Moses that it was about time to make a Sun and Moon."

Here Mr. Ingersoll again betrays a good deal of unfairness. Scientists of the highest standing agree that the Earth was covered with the most luxuriant vegetation, long before any ray of the Sun could penetrate the dense atmosphere which surrounded our globe. When the heat of the Sun's rays was not yet felt upon the Earth, her own internal heat sufficed to call forth a tropical flora, even where now everlasting ice and snow reign.

---

6. Naturforschung und Bibel.
7. Some Mistakes of Moses, pp. 68, 72.

Louis Figuier observes[8] on this subject : "It is a remarkable circumstance that conditions of equable and warm climate, combined with humidity, do not seem to have been limited to any one part of the globe, but the temperature of the whole globe seems to have been nearly the same in very different latitudes. From the Equatorial regions up to Melville Island, in the Artic Ocean, where, in our day, eternal frost prevails,—from Spitzbergen to the centre of Africa, the carboniferous flora is identically the same. When nearly the same plants are found in Greenland and Guinea ; when the same species, now extinct, are met with of equal development at the equator and the pole, we cannot but admit that at this epoch the temperature of our globe was nearly alike everywhere. There seems to have been then only one climate over the whole globe. * * * Whence then proceeded this general superficial warmth, which we now regard with so much surprise ?—It was a consequence of the greater or nearer influence of the interior of the globe. The Earth was still so hot in itself, that the heat which reached it from the Sun may have been inappreciable. * * * No flower yet adorned the foliage or varied the tints of the forests. Eternal verdure clothed the branches of the Ferns, the Lycopods, and Equiseta, which composed, to a great extent, the vegetation of the age. * * * No food appeared fit for nourishment. * * * A damp atmosphere, of an equal rather than an intense heat like that of the tropics, a soft light veiled by permanent fogs,

---

8. The World before the Deluge. pp. 133, 137, 151.

were favourable to the growth of this peculiar vegetation, of which we search in vain for anything strictly analogous in our own days. The nearest approach to the climate and vegetation proper to the geological period which now occupies our attention, would probably be found in certain islands, or on the littoral of the Pacific Ocean—the island of Chloe, for example, where it rains during 300 days of the year, *and where the light of the Sun is shut out by perpetual fogs.*"

Dr. Lorinser remarks[9] on the appearance of plants before the Sun: "It is most probable that formerly the condition of the atmosphere was such that what we now call a clear sky, was then impossible. The air was filled with carbonic acid, and the great internal heat of the Earth continually caused dense vapors to rise from the waters, which formed such masses of clouds and fog, that the sidereal globes (Sun, etc.,) remained veiled to the surface of the Earth. The coal period caused an important change: the great absorption of carbonic acid by the luxuriant plants of the coal period, helped greatly to clear the sky. Perhaps this was even the main object of this grand vegetation. Consequently, the work of the third Day was in casual connection with the appearance of the Sun and the other sidereal bodies. Therefore, the appearance of these could not be mentioned in a chronologically more appropriate place, than immediately after the creation of the plants."

We may add, although the creation of the plants is ascribed to the third Day, we are not obliged to

---

9. Geologie und Paleontologie.

infer from this, that no new species of plants appeared after the third Day. We may say, with Dr. Lucken :[10] "The Bible mentions only the first appearance (Grundlegung) of every new element in creation. What the created, natural forces, inherent in these elements, may have changed or transformed, within their fixed limits, is not the object of the doctrine of Creation to reveal. * * * Since the plants, more than other organisms, are completely dependent on the soil and climate, we may expect only a gradual transition to our present flora, namely, as soil, climate, and air gradually arrived at their present condition."

The same, substantially, may be said of the creation of the animals, mentioned on the fifth and sixth Days.

These explanations will suffice to show that, in spite of seeming contradictions, real harmony exists between the Mosaic record and the results of modern scientific investigations.

Both Moses and modern scientists teach : 1. That the Earth was once "void and empty," without plants, animals, or man, and covered with darkness,—dense vapors arising from the hot surface of our globe. 2. Then, as Prof. Winchell observes, "the Sun in the morning sent a lurid ray through the dense atmosphere, and at night sank (apparently, looked at from the surface of our globe) into the smoke that ascended from a burning world;" or as Moses teaches,—God divided the light from the darkness,—day and night began to alternate. 3. Next, the expanse, or firmament, which

---

10. Die Stiftungsurkunde des Menschengeschlechts, Freiburg, 1876.

separates the waters on the Earth from the waters in the clouds, became discernable. 4. After this, dry land emerged from the warm oceans that covered the Earth; and finally the Sun, Moon, and Stars, in all their brightness, became visible from the surface of the Earth; the dense, dark atmosphere having gradually disappeared. 5. After the surface of the earth was sufficiently cooled and prepared, the first organisms appeared,—plants before animals.

The order in which animals appeared, was, according to Moses: First, the animals living in water, and the birds; next, the animals living on dry land; and finally, man. The same is taught by Modern Science. Prof. Winchell observes,[11] on the succession of animal life on Earth: "Life has presented itself * * * in a succession of *dominent ideas*, each in its own way expressing itself in more than one organic type. Thus, in the reign of *reptiles*, the reptilian idea was dominant, and we find it invading the structure of the contemporaneous *fishes*. Afterward the *ovian* or *ornithic* (*bird-like*) idea became dominant, and the reptiles were endowed with wings. Still later the *mammalian* idea became dominant."

Here we find the same order in the succession of animal types given, which we find indicated in the Mosaic Record.—Thus we see that there exists an admirable harmony between Moses and Modern Science. —Wherein Moses differs from the latter, may be easily accounted for, if we consider the different stand-points

11. Sketches of Creation.

from which Moses and Modern Science view the same work of creation.—The object of Modern Science is to give an exact *scientific* account of the relative co-existence or succession of plants and animals upon the Earth. —Moses, on the other hand, whose main and only object here was, to teach mankind that *God is the Creator of* ALL *organic beings*, considered it sufficient, for his purpose, to mention the different acts of the creation of *all* kinds of plants and animals.

V. BIOLOGY.

1. THE APPEARANCE OF ORGANIC AND LIVING BEINGS ON THE EARTH.

MR. INGERSOLL observes: "There must have been a time when plants and animals did not exist upon this globe."

Certainly, on this point all agree — Moses and modern scientists. Moses mentions expressly that the Earth was once "void and empty," Genesis 1, 2 ;— that is, without plants that adorned, or living beings that inhabited it ; and scientific men say that not only the Earth but the whole visible universe was once in an intensely hot state, which would have made all life, we are acquainted with, impossible.—Therefore, also the theory of Thompson and Helmholz,[1] that the germs of the first living or organic beings came with meteors from other parts of the universe upon our Earth, does

---

1. Naturforschnng und Bibel, von C. Guettler, p. 133.

not bring us any nearer to the solution of the question as to the origin of life. Even if we would admit the theory just mentioned—what we are by no means inclined to do—the question would still remain to be answered: "How did the germs of living beings get on those meteors, the elementary constituents of which must, according to the theory of Laplace, once have been in such a state of tenuity and incandescence as to make the existence of all germs of life, Science knows anything of, impossible?"

Where, then, did the first plants and animals come from?—Infidels assume, as a matter of course, that they must have in some way sprung from pre-existing inorganic matter alone; else they would be compelled to admit a Creator. This idea would, of course, be shocking to them.—Mr. Ingersoll in *Some Mistakes of Moses*, declares: "This is so grossly improbable, so at variance with the experience and observation of mankind (as if mankind had been present at the first acts of creation!), that it cannot be adopted without abandoning forever the basis of scientific thought and action."—Mr. Ingersoll assumes that the denial of the Creator is that "basis."—Yet on this point he is at variance with men greater than himself. As Mr. Ingersoll, also other unbelievers assume that organic and living beings *must* have sprung—they do not know how—from inroganic matter without any outside influence;—for else they would have to admit a Creator!

Infidels, moreover, go on assuming that the first living organisms commenced and continued to develope themselves, of course, without the help of a Creator, until finally the present world was evolved, with all its plants, animals, and man.

Moses took the liberty to differ on these points with our infidels; for he not only insinuates, but expressly teaches that God, the Creator, had something to do with bringing into existence the plants, animals, and finally man. We read in the book of Genesis: "God said: 'Let the Earth bring forth the green herb, * * * the fruit tree, etc.'" Now, who is right: Moses, or Mr. Ingersoll and other infidels?—What does true Science,—which is not to be confounded with the bold assertions of such infidels as Ingersoll, Buechner, Haeckel, etc., teach on this subject?

Although the inmost nature of inorganic matter and force, as all profound scientists confess, is still as much a mystery as ever, yet that much, at least, Modern Science has established, that there is no proof that organized life of any kind ever sprang from inorganic matter alone. Before the discovery of the microscope, it was assumed by some naturalists, that lower kinds of plants and animals may be produced from inorganic matter. The authority of Aristotle induced even such great men as St. Augustin, Peter Lombard, St. Thomas of Aquin, and the Scholastics of the Middle Ages generally, to adopt this view;—which in itself cannot be said to be anti-Christian. For, as Dr. Guettler[2] remarks: "The Creator loses nothing of His dignity and perfection, if we assume that He has given

2. Natur forschungund Bibel, p. 135.

to (inorganic) matter the power to produce, under certain circumstances, living beings."

Yet, careful modern scientific experiments have proved beyond a doubt, that even the most minute organic or living beings, whether plants or animals, are never produced from inorganic matter, but always spring from some pre-existing seeds or germs of their own kind. The famous French scientist Pasteur has proved beyond any reasonable doubt: first, that the atmosphere is filled with minute seeds or germs of microscopic plants and animals; secondly, that if the necessary precautions be taken, to exclude these germs or seeds from an apparatus, filled with matter in which such organisms otherwise grow, they will not appear; but if the ordinary atmosphere is permitted, only for a moment, to come in contact with the matter in the apparatus, within a day or two some of the microscopic plants or animals will make their appearance.—Even up to 1873 the English scientist Bastian imagined to have proved that such microscopic organisms could be produced from inorganic matter;—for some had appeared in carefully closed up matter which had been subjected to a heat of 212 degrees F.—But other English scientists, W. H. Dallinger and Dr. Drysdale, exposed Mr. Bastian's error by proving by experiments, that some microscopic organisms or seeds are not destroyed by a heat of even about 390 degrees.[3]

Careful experiments have convinced modern scientists, that, as far as observation, aided by the most perfect

---

3. Dr. Fr. Lorinser: Geologie und Palaeontologie, p. 301.

microscopes, reaches, not a solitary instance can be discovered, to prove that any living or organic being, plant or animal, can be produced from inorganic matter alone.

Now, since all scientific results tend to prove that the same laws of nature have always existed—as long as matter has existed—it is a mere subterfuge of our infidels to say :—"Although *now* no such instances can be found, yet in remote geological ages it *may* have happened that life sprang from purely inorganic matter." —According to the fixed laws of nature, what is impossible to-day, must have been impossible always since nature, as it is, exists.

The infidel scientist Haeckel[5] imagined to have, found in the so-called Bathybius, a jelly-like substance, discovered by Huxley, "the missing link" between inorganic matter and living organisms. But some suspect that this Bathybius, which was said to have been found at the bottom of the ocean, was nothing but jelly-like gypsum, and no living being at all. Asa Gray, the eminent American naturalist, observes :[4] "This living matter—of which Bathybius, *if there be a Bathybius*, or if it be anything more than protoplasm of sponges, is one example—is said to have nothing more than molecular structure. It would be safer to say that the microscope has as yet revealed no organic structure."

No doubt, there are most simple jelly-like living beings, called moners or amoeba, living in water, of

5. Natuerliche Schoepfungs Geschichte, 4 edit., p. 165.
4. Natural Science and Religion, New York, 1880, p. 15.

which Prof. Winchell[6] says : "There is no stomach, no liver, no heart, no breathing organ, no head, no feet —in short, this animal is destitute of organs, except as it employs the whole body for every purpose. Whenever it seizes its food, it extemporizes an arm for the work. Whenever it eats, it must extemporize a stomach. It seizes, it eats, it digests, it breathes with the whole body." More simple living beings cannot be imagined.

Yet to come to the main point at issue : Are these microscopic, jelly-like living beings produced from inorganic matter ?—No, there is no proof for such an assumption, and Haeckel[7] himself teaches that they are propagated or multiplied by self-division (Selbsttheilung); as, for instance, some worms may be divided, and each part will continue to live by itself.

Thus we see that, according to the present known facts and laws of nature, even the most imperfect living beings are produced only from organic beings of their kind, and not from any inorganic matter alone. Since, then, as all observation teaches, the laws of nature have never changed, we may safely conclude that this was always the case ; however infidel scientists, who, like Haeckel,[8] would like to explain nature without admitting a Creator, may struggle against this inevitable conclusion.

Thus we see that the inspired writer Moses is in perfect harmony with Modern Science in teaching that organic and living beings did not spring from inorganic matter alone, but appeared *at the word of God.*

6. Sketches of Creation, pp. 70, 71.
7. Natuerliche Schoepfungsgeschichte, Berlin, 1873, p. 167.
8. L. c., pp. 302, 303.

Mr. Haeckel [9] has yet a straw to cling to in his struggle against a Creative Power, interfering with the origin of organic and living beings.—He admits that there exists no instance to show that any organic or living being has ever been produced from inorganic matter alone; yet he hopes that *in some distant future* Chemistry will succeed in producing some. He claims that Modern Chemistry has succeeded in producing "organic" products, as alcohol, formic acid (Ameisensæure), etc. Therefore he expects, that sooner or later also other "organic" bodies will be produced artificially; whereby, of course, "the deep chasm between organic and inorganic bodies" would be bridged over.

To this the great chemist Justus von Liebig [10] replies: "Never will Chemistry be able to produce an eye, a hair, or a leaf." [10] "What those charlatans (Dilletanten) call organic products (organische Verbindungen) are no such at all, but chemical ones. But never will Chemistry succeed in producing a cell, a muscle-fibre, a nerve, or, with one word, any part of a truly organic organism, endowed with vital properties. * * * Inorganic forces will always create only something inorganic." [11] The authority of such a profound chemist, as Justus von Liebig, outweighs by far the unfounded assumptions of prejudiced charlatans, of whom the great chemist remarks: "The same charlatans in natural science, who know not what a fever or an inflamation is, or how

---

9. L. c., pp. 203, 204.

10. Chemische Briefe, Leipzig und Heidelberg, 1865, p. 14.

11. L. c., pp. 205-6.

the blood is produced, or to what purpose the bile is, * * * want to make the credulous public believe that they can explain the origin of thoughts, the nature and the essence of the human spirit." [12]

2. MODERN SCIENTIFIC VIEWS ON MATTER AND LIFE.

The relations existing between matter and life being one of the principal battle-grounds between modern materialism and Christian philosophy, let us examine this subject more fully.

In the preceding article we have seen that, as exact and reliable experiments prove, not even the lowest, or most imperfect, organic or living beings can be produced by or from inorganic matter alone. "This great principle," as Francis Emily White, M. D., remarks,[1] "was, indeed, recognized by Harvey, and first expressed in his famous aphorism, 'Omne vivum ex ovo'—(every living being comes from an egg, or germ)."—And of this "ovum" we may say with Ralph Waldo Emerson:[2] "All we know of the egg, from each successive discovery, is, *another vesicle;* and if, after five hundred years you get a better observer or a better glass, he finds, within the last observed, another."

Although, then, organic and living beings cannot be produced by lifeless, inorganic matter,—yet there exist certain relations between matter and life.

Let us see what Modern Science says on this subject.

12. L. c., p. 207.
1. The Popular Science Monthly, March, 1884.
2. The Conduct of Life, Boston, 1884, p. 19.

Mr. Huxley, in his famous lecture "On the Physical Basis of Life," remarks: "There is some one kind of matter which is common to all living beings. * * *

Beast and fowl, reptile and fish, mollusk, worm, and polype, are all composed of structural units of the same character, namely, masses of protoplasm with a nucleus. * * * What has been said of the animal world is no less true of plants. * * * All the forms of protoplasm which have yet been examined contain the four elements, carbon, hydrogen, oxygen, and nitrogen, in very complex union." Asa Gray [3] endorses this doctrine, and adds: "The statement that 'protoplasm is the physical basis of life' must be accepted as true. As Professor Allman puts it, 'wherever there is life, from its lowest to its highest manifestations, there is protoplasm; wherever there is protoplasm, there too is life, or has been." How do living organisms, as far as scientific observations can learn, develop from protoplasm?—This question is answered by Professor Asa Gray, [4] as follows: "Each ordinary plant or animal, begins as one cell, which is then the simple individual. This in growth divides itself into two, these two into four, these into sixteen, and so on, building up the structure. * * * Higher in the scale structures are built up, what were individuals, become parts or organs. * * *; then the life of the cells is their own no less, but their individuality blends in the common life of the aggregate. By increasing complexity of organization,

---

3. Natural Science and Religion, pp. 12–13.
4. Natural Science and Religion, pp. 30–34.

with increasing subordination of parts and specialization of office, the highest plants and animals are composed." But does this cell theory explain the mystery of life? Professor Gray answers : "It is an illusion to fancy that the mystery of life is less in an amoeba or a blood-corpuscle than in a man."

On these points, then, modern scientists agree : that protoplasm is the physical basis of the life of all organic beings, and that all organic structures are made up of cells.

Another kind of facts which prove the intimate connection between matter and life in all organic beings, is the so-called correlation between vital and physical or chemical forces.

Professor Barker of Yale College remarks :[5] "The important fact must be fully recognized that in living beings we have to do with no new elementary forms of matter. Precisely the same atoms which build up the inorganic fabric, compose the organic."

Observation will easily convince any one that the material elements drawn by the plant from the soil and the atmosphere, are the food by which the animal and human body, the bones, muscles, nerves, etc., are built up. Prof. Barker continues : "The heat produced in the body is precisely such as would be set free by the combustion of this food outside of it. * * * No doubt can be entertained, that the actual energy of the muscle is simply the converted potential energy of the carbon of the food."

---

5. The Correlation of Vital and Physical Forces.

Also the nerves and brain, the instruments of the human soul, are composed of matter, and are subject to physical and chemical influences. A diseased state of the brain-matter may disturb even the noblest faculty of man—his intelligence.

Thus we see that matter—protoplasm—is not only the physical basis of organic life, but also that physical or chemical forces are intimately connected with vital phenomena of even the highest kind; so that the Editor of "The Popular Science Monthly"[6] feels justified in remarking: "In dealing with abnormal mental manifestations, as in the numerous forms of insanity and the various grades of feeble-mindedness, or with the psychological effects of stimulants and narcotics, or with the development and decline of the mental powers, or with the effects of mental overwork and exhaustion, it is now admitted to be indispensable to start from the nervous system, and to regard mental manifestations as conditioned by its properties and laws."

From all this we see how matter and physical or chemical forces are most intimately connected with the noblest phenomena of life.—Prof. Joseph Le Conte, of the University of California, observed in a lecture:[7] "Vital force; whence is it derived? What is its relation to the other forces of nature? The answer of Modern Science to these questions is: It is derived from the lower forces of nature, * * * it is correlated with chemical and physical forces. * * * There are four

7. Popular Science Monthly, December, 1873, p. 156-170.
6. November, 1873, p. 110.

planes of material existence (the animal, vegetable, and mineral kingdoms, and the elements), which may be represented as raised one above another. * * * Now it is a remarkable fact that there is a special force, whose function it is to raise matter from each plane to the plane above. * * * Plants cannot feed upon elements, but only on chemical compounds: animals cannot feed on minerals, but only on vegetables. * * * Now, there are also four planes of force similarly related to each other, viz., physical force, chemical force, vitality and will (sensation?). The change from one grade to another, as from physical to chemical, or from chemical to vital, is not, as far as we can see, by sliding scale, but suddenly. The groups of phenomena which we call physical, chemical, vital, animal, rational, and moral, do not merge into each other by insensible gradations."

Thus we see the intimate connection between matter and life, and between physical, or chemical, and vital forces.—But has Modern Science with all its ingenious and exact observations of material phenomena discovered the inmost nature of life, of sensation, of intelligence, of will?

We see the leaves, twigs, branches, and the stem of a great tree,—but its roots are hidden to our eyes,—so we also behold the manifold phenomena of life,—but their main-spring, their cause, is, as we shall see in the next article, still as mysterious to us, as it was in the days of Aristotle and Confucius; although some "advanced thinkers" talk as if they had fathomed all the mysteries of the universe.

3. MYSTERIES OF NATURE.

What is matter ? What is force ? What is vitality ?
What is sensation ? What is intelligence ? What is
will ?

These are questions to which all modern scientists, in
spite of all scientific progress, must answer :—*Ignoramus*,—We know not, and shall, as mortals, never
know ; for neither the microscope, nor the telescope,
nor the spectrum-analysis, nor any imaginable instrument is likely to ever throw any light upon these
mysteries of nature. Let us see what reflecting
scientists say of these questions :

What is matter? Henry Hobart Bates, M. A.,
answers :[1] "The nature of matter is still almost as
unknown to us in its essence as it was to the ancients,
since in its minute structures it lies far below the range
of senses, or of instrumental appliances, and therefore
beyond that direct experimental field so necessary in
furnishing primary conceptions to the mind. * * *
The great trouble about matter is to find out how much
of it and what in it is material. Strange to say, there
is nothing on which philosophers are less agreed."

The same is true of the nature of force, which, still
more than matter, eludes instrumental appliances.
Emil Du Bois-Reymond, in his famous address on " The
Limits of our Knowledge of Nature," delivered at the
forty-fifth Congress of German Naturalists and Physicians at Leipzig, August 14, 1872, remarked : " In
our endeavor to analyse the physical world, we start out

---

1. The Popular Science Monthly, April, 1883, pp. 788-9.

from the divisibility of matter, the parts being to our eyes something simpler and more primitive than the whole. When in thought we carry on this division of matter ad infinitum * * * we meet with no obstacle in the process. But we make no advance whatever toward an understanding of things, since we, in fact, carry over into the region of the minute and the invisible the concepts we obtained in the region of the gross and the visible. * * * The ancient Ionian physical philosophers were no more helpless than we in the presence of this difficulty. The natural sciences, with all the progress they have made, have availed naught against it, nor will their future progress be of any greater effect. * * * For two thousand years, despite all the advances made by natural science, mankind has made no substantial progress toward the understanding of matter and force, any more than toward the understanding of mental activity from its material conditions. And so will it ever be. * * * With regard to the enigma of the physical world the investigator of Nature has long been want to utter his 'Ignoramus' (we know not) with manly resignation. * * * As regards the enigma what matter and force are, and how they are to be conceived, he must resign himself once for all to the far more difficult confession —'Ignorabimus'—(We shall never know)."

Still more mysterious and incomprehensible, than matter and physical force, seem to be—vitality, sensation, intelligence, and will.

Speaking of life, the same scientist remarks in the address mentioned before : "There comes in, at some point in the development of life upon the Earth, which we cannot ascertain — something new and extraordinary; something incomprehensible, again, as was the case with the essence of matter and force * * * This other incomprehensible is consciousness * * * I use the term "consciousness" designedly, the question here being only as to the fact of an intellectual phenomenon, of any kind whatsoever, even of the lowest grade * * * Just as the most powerful and best developed muscular performance of man or animal is in fact no more obscure than the simple contraction of a single muscle — as the single secretory cell involves the whole system of secretion—so the most exalted mental activity is no more (?) incompresensible in its material conditions, than is the first grade of consciousness, i. e. sensation. With the first awakening of pleasure or pain, experienced on Earth by some creature of the simplest structure, appeared that impassible gulf, and then the world became doubly incomprehensible * * * * The highest grade of knowledge (of the brain) we can ever expect to have— discloses to us nothing but matter in motion. But we cannot, by means of any imaginable movement of material particles, bridge over the chasm between the conscious and unconscious * * * What conceivable connection subsists between definite movements of definite atoms in my brain, on the one hand, and on the other hand such (for me) primordial, indefinable, undeniable facts as these: I feel pain or pleasure; I experience a sweet taste, or smell a rose, or hear an organ, or see something red — and the immediately consequent certainty. 'Therefore I exist ?' It is absolutely and forever inconceivable that a number of atoms of carbon, hydrogen, nitrogen, oxigen, etc., should not be indifferent as to their own position and motion, past, present or future. It is utterly inconceiveable how consciousness should result from their joint action. If their respective positions and their motion were not indifferent to them, they would have to be regarded as each possessed of a consciousness of its own, and as so many monads. But this would not explain consciousness in general, nor would it in the least assist us in understanding the unitary consciousness of the individual."

These words of the famous infidel scientist Du Bois-Reymond show that Modern Science neither can now, nor has it any prospect of ever being able to explain what that mysterious principle is which causes consciousness, or sensation, even in a worm or a microscopic animalcule. Much less can Modern Science explain what human intelligence, or reason, is. The great German chemist L. Justus von Liebig speaks with just indignation of those would-be scientific charlatans who are ignorant of what a fever or an inflamation is, and yet impudent enough to endeavor to make the credulous public believe that they are able to explain — "the origin of thoughts, and the nature and essence of the human spirit." [2]

Indeed, Modern Science is not able to explain the nature of even the lowest form of life—vegetation. Modern Science by means of the most perfect microscopes has discovered, as was mentioned before, that all plants

---

2. Chemische Briefe. 23. Brief.

are made up by great numbers of so-called cells.   Prof.
Ferdinand Cohn of Breslau remarks in an article on "The
Cell-State :"[3] "We owe it to the microscope that, where
the naked eye perceives only uniform masses (in plant-
life), we can now distinguish a wonderful diversity of
beautiful tissues ; and that where a rigid stillness seems
to prevail, a fullness of life-processes quite incomprehensible to us is concealed.   The microscope shows us
in the plant, which was able to give the naked eye
only obscure signs of its inner life, a highly organized
state-life of restless development and renewal."
Having compared the plant to an organized state in
which all citizens—the plant-cells—harmoniously work
together for the common welfare, Prof. Cohn continues :
" We have represented the citizen of this state, the
plant-cell, as an exceedingly simply formed being : it
consists of a round body of soft, slimy substance, like
a sack, the interior of which is filled with a watery
juice. The soft substance * * * is called protoplasm ; it is the most important matter in all nature, for
it alone is the bearer of life. * * * A continuous
formation and transformation, origin and decay, a
constant change of matter, is going on in every cell ;
reception and assimilation of food, inspiration and
expiration ; certain atoms which have become of no use
for purposes of life are cast aside, others are taken up
from without in their places. * * * Evidently not
solid substances are appropriated, for we know that the
cell is encased in a perfectly closed envelope ; but

---

3. The Popular Science Monthly, December, 1882, p. 174-188.

liquid and gaseous foods can be easily absorbed. Although the most perfect microscopes have never made any holes visible in the cell-envelope, there is not the slightest doubt that this envelope is porous, like a fungus, but that its pores are infinitely finer."

Now, that is about all the explanation Modern Science can—and probably ever can — give on plant-life; the most perfect microscope can reveal no more.—Every plant, then, consists of cells, of which every one is living, and all work harmoniously together for the welfare of the whole plant, to produce roots, stem, leaves, blossoms, and fruit. But what is that mysterious principle which causes each plant cell to live and work, —and which compells all plant-cells to work harmoniously together according to a certain plan? No modern scientist can tell, and probably no one will ever be able to tell; for that myterious principle eludes the most perfect microscope and the most subtle chemical analysis.

Before the mystery of life, Modern Science stands as ignorant as the Greek philosophers centuries before Christ. Every reflecting scientist will agree with Mr. Burmeister when he says : " What the principle of life (Lebenskraft, vitality) is, we know as little, as what force itself is; and we must therefore remain satisfied with the meagre explanation that it is the cause of the all phenomena of matter."[4] That's just what we would like to know, what exactly that *cause* is. But as reflecting scientists admit—science cannot, and probably never can, tell.

---

4. Das Exacte Wissen der Naturforscher, von D. von Schuetz, p. 7

Thus we see that Modern Science, in spite of its marvellous progress in many directions, is still surrounded by impenetrable mysteries, which are indeed the very sources of all the phenomena of the visible world. — What is matter ? What force ? What life ? What sensation ? What intelligence ? What will ? — The answer of Modern Science to these questions is contained in the words of Du Bois-Reymond : "Ignoramus — Ignorabimus ;" we know not—and in this life never shall know ; — and yet all phenomena of the visible world have their source in matter, life, sensation, etc.—

We may compare the whole visible universe to a gigantic tree. We see its leaves, branches, and stem ; — but the roots — from which these spring, — are hidden from our eyes. — We may compare this visible universe to a mighty stream flowing in quiet majesty through the plains of time and space. We see the numberless wavelets on its surface, as they glide by, but we see neither the bottom of the stream, nor the source from which it springs.

Natural sciences which are based on experiments and observation alone, can never explain to us the mysteries of matter, force, life, or intelligence. Indeed, no branch of human sciences can lift perfectly the veil that conceals those mysteries from our view.

Yet, there is a science which penetrates a little further into these mysteries than the strictly so-called natural or experimental sciences. The science to which I allude, is Christian Philosophy — which is built up on the two solid foundations of Human Reason and Divine Revelation.

Let us see what Christian Philosophy teaches on the mysteries of the visible universe.

4. CHRISTIAN PHILOSOPHY ON MATTER AND LIFE.

As we have seen in the preceding article, deeply hidden mysteries underlie the whole visible creation; — mysteries, the solution of which Modern Science is giving up in despair. No man ever has known fully, and no mortal ever will know, what the intrinsic natures of matter, force, vitality, sensation, intelligence, or will, are. — Let us therefore, without uselessly troubling ourselves about the unknowable, be content with what we *can* know on the subject.

Christian Philosophy has penetrated as far, as man will ever be able to do, into those mysterious recesses of nature, — into which neither the telescope, nor the microscope, nor the spectrum-analysis can penetrate.—Here is a realm of hidden truths that can not be touched with the hand, or seen with the eye, but must be reached by reason—or the intellect, which does not stop at the visible phenomena, but penetrates beyond them, by reading the nature of the causes in their effects. Therefore this faculty, with which man alone in the visible creation has been endowed by the Creator, has been called "intellect",—from "intus", or "inter" and "legere" — to read within, or between; — because by means of this faculty man can, to some extent, read what is concealed behind, or within, or between visible phenomena.

Now let us see what Christian Philosophy teaches on the subjects of matter and life.

St. Augustine[1] remarks : "Two things Thou, O God,
hast created : the one quite near to Thee (the angels);
the other quite near to nothing (matter). The one has
nothing above itself, but Thee ; the other has nothing
below itself, but the nothing." St. Bonaventure expresses the same thought with these words - "It was becoming that God created not only a substance far distant
from Himself, namely the corporeal nature, but also a
substance near to Himself ; and this is the intellectual
substance (the angels)." — In man, the moral center of
creation, we find both natures, or substances, perfectly
united — matter or body — and spirit.—Divine Revelation teaches that higher in the scale of creation than
man, are numberless angels of different degrees of perfection, whom man resembles as to his spiritual, intelligent soul ;—and below man, in the scale of creation, we
find numberless gradations of beings,—which have more
or less similarity with man. On one point all corporeal
beings, from a particle of dust to the most perfect animal, have a similarity to man : in all we find two distinct
yet intimately united principles of their being — a material and an immaterial one. The former is called by
Christian Philosophy "materia", or matter,— the latter,
"forma," or the formative principle. The Jesuit Father
M. Liberatore[3] says on the essential composition of
bodies : "The reality * * * from which extension results, is best called materia (matter); the extensionless
force is best called forma."

---

1. Confessionis, XII, 7.
2. Brevilognium II, 6.
3. Institutiones Philosophicae, Romae, 1861, p. 460.

Now, these two principles we find in every body : in a particle of dust as well as in minerals, plants, animals, and man ; — a material, passive, principle from which arises the extension of the bodies, and an non-material, active principle which gives bodies their peculiar nature, or mode of existence and operation.

In visible nature, we find these two principles united everywhere ; yet the more perfect in the scale of creation beings are, the more the "forma," or active principle, becomes superior to or independent of matter. — In minerals, we find the "materia" and "forma" in seemingly perfect interdependence.—In plants, the active principle, or "forma", shows itself more plainly by marshalling the particles of matter according to certain ideas or plans.—In animals, the non-material, active, principle not only marshals the material particles within the body of the animal, but also moves the whole body, receives sensual impressions, and causes the most complicated reflex actions. In man, the active principle becomes, by abstraction and reflexion, perfectly conscious of its existence, it perceives the nature of things surrounding him, etc.—And in angels finally, as we know from Revelation, we find the active principle, or "form," in its purity — without any material admixture.

After these general remarks, let us briefly consider the differences between inorganic matter and plants,—between plants and animals,—and between animals and man.—Now-a-days especially it is necessary for intelligent Christians to have a clear idea of these differences, because infidel materialists endeavor by all means, if this were possible, to explain all these differences away,

—and to explain the whole visible universe, man included, on the shallow assumption that all living beings are merely complicated formations gradually developed from inorganic, material molecules without any intervention of a wise and powerful Creator.

Thus, for instance, Prof. John Tyndall, in his famous Inaugural Address before the British Association,[4] remarked : "Trace the line of life backward, and see it approaching more and more to what we call the purely physical condition. We reach at length those organisms which I have compared to drops of oil suspended in a mixture of alcohol and matter. We reach the *protogenes* of Haeckel, in which we have 'a type distinguishable from a fragment of albumen only by its finely granular character.' * * * I prolong the vision backward across the boundary of experimental evidence, and discern in * * * matter * * * the promise and potency of every form and quality of life."

Such, in substance, is the view of some infidel scientists who, to get rid of a Creator, let their imagination escape with their reason "across the boundary of experimental evidence."—Cool-headed scientists, on the contrary, as we have seen in a previous article, maintain that, as far as "experimental evidence" reaches, no organism whatever, plant or animal, however small, has yet been known to have originated directly from inorganic matter ; in every observed case, it was an organic germ, or ovum, from which the organic being sprang.

---

4. The Popular Science Monthly, October, 1874, p. 681.

Not only inorganic matter, but also organizing forces are necessary for the production of any organic being. The nature of organic beings does not depend so much on the material substrata, as on the formative forces operating on these.

As we have seen before, the so-called " protoplasm," composed of carbon, hydrogen, oxygen, and nitrogen, is " the physical basis " of all plant and animal life,—and is physiologically and structurally the same in both plants and animals.[5] Moreover, each ordinary plant or animal begins as one cell, which multiplies, building up the whole plant or animal structure.[6] From this we see that the material substratum is essentially the same in plants and animals.—But what causes the countless differences between plants and animals ?—Not the material substrata which are essentially the same in all, but the different " formal " or formative principles which are united with these material substrata.

On this point Christian Philosophy and experimental evidence agree perfectly ;—and, indeed, Modern Science is on this point not a step ahead of Christian Philosophy as taught in the days of St. Thomas Aquinas. Let us then see what Christian Philosophy teaches concerning matter and life,—and how its views are corroborated by modern scientists and philosophers.

Father Liberatore,[7] one of the most profound modern philosophers, following the principles of St. Thomas, teaches : " In living beings there is a certain essential

5. Natural Science and Religion, by Asa Gray, p. 13.

6. Ibid., pp. 30, 31.

7. Institutiones Philosophicae, Rome, 1861, p. 501.

principle, quite distinct from physical and chemical forces." It is owing to this principle, that plants grow and propagate themselves; it is owing to this principle, that animals have sensations, etc.; it is owing to this principle, that man reasons, etc. But this principle is not the same in plants, in animals, and in man, as the different effects prove. In plants, it causes only vegetative life, in animals, the vegetative and sensitive life,—in man, finally, besides causing the vegetative and sensitive life, it is also the source of reason,—that faculty which elevates man beyond the material world, and places him in the realm of the spirits.

This is, in substance, the view of Christian Philosophy on matter and life. Let us see what modern scientists say on this same subject.

Dr. Julius Adolph Stoeckhardt,[8] one of the most famous German chemists, observes: "As long as a plant or an animal lives, the chemical processes stand under the guardianship of a superior, mysterious power, which is called vitality (Lebenskraft), and are forced by it to gather the materials for the structure of the plant or animal body. Vitality is, as it were, the architect that makes the plan of the building, and sees to it that the necessary materials are procured by chemical processes, and disposed according to its will. * * * When life ceases in a plant or an animal, the chemical processes gain the upper hand, and they alone, as scavengers (Todtengraeber) of nature, fulfill the old saying: 'What is of earth, shall again return to earth.' "

8. Die Schule der Chemie, No. 3.
9. Ibid., No. 546.

Again [9] Dr. Stoeckhardt remarks on the life of plants : "An Incomprehensible Wisdom has endowed the seed with a force which causes the seed, placed in moist soil, to germinate, to grow up as a plant, to bring-forth leaves, blossoms, and seeds (or fruit). * * * If they (the plants) have produced seeds, that is, new bodies endowed with vitality, they have fulfilled their object, and their course goes downward towards decomposition. Whether they come to this end after a short summer, or after hundreds of years, makes no material difference. The breath of God, which calls forth these changes, the phenomena of life, in the world of plants, is perfectly unknown to us as to its nature (or essence). It has indeed been called vitality (Lebenskraft) ; yet thereby we have not come to a clearer understanding of its being. Its operations are conducted in such a mysterious manner, that it seems as if the yearning (Ahnen) of the investigating mind of man shall, in this respect, here below, never be changed into clear vision."

Another famous German chemist, Justus von Liebig,[10] also insists on the necessity of distinguishing chemical processes from the effects of the vital principle (Lebenskraft), in plants and animals. He declares that in organic, living beings a formative (formbildendes) principle, a ruling idea, works in and with the chemical forces.

These two great chemists but assert what unprejudiced scientists generally admit—and what Christian Philos-

10. Chemische Briefe.

ophy has taught for centuries : that in living organisms a life-giving principle—superior to the known chemical and physical forces—must be admitted.

Some infidel scientists who would fain explain all phenomena according to their preconceived theory of atomic materialism, are inclined to deny this assumption. Prof. John Tyndall even imagines to have found the connecting link between mere inorganic matter and living plants—in crystals.—In an article on " Crystalline and Molecular Forces," [11] he says : " How have these crystals been built up ? *   *   *   Without crossing the boundary of experience, we can make no attempt to answer this question. * * *   From the processes of crystallization * * * you pass by almost imperceptible gradations to the lowest vegetable organisms, and from these through higher ones up to the highest."

Mr. Tyndall admits that he cannot explain the formation of crystals " without crossing the boundary of experience ; " he imagines, nevertheless, that one may pass from the process of crystalization gradually to vegetable organisms.   In imagination only,—Mr. Tyndall should have added.

No doubt, the processes of crystallization prove that the Creator has endowed certain matters, as sugar, salt, sulphur, lead, etc., with a mysterious force which compels, under certain circumstances, their molecules to form, with mathematical precision, certain configurations. Yet, the most perfect crystal is as much different from even the lowest vegetable form, as a sculptured horse from a living one.

---

11. The Popular Science Monthly, January, 1875, pp. 257-68.

Every crystal is formed under certain circumstances, —and then, if not disturbed by outside influences, remains unchangeable for any length of time. The stamp of stability—of death, is imprinted on it. On the other hand, we find motion from within — life —in the very lowest of vegetable forms; they all germinate, grow, and reproduce their species *ad indefinitum*. That's more than any crystal can do.

Having considered the boundary line betweeen lifeless matter, and plants, let us next see what separates the vegetable from the animal kingdom.

Many modern scientists look more to the material elements found in plants and animals than to the so-called "forma," or active principle, which causes the differences between vegetable and animal organisms. Prof. Asa Gray, for instance, remarks:[12] "We cannot conceive anything more characteristic of a vegetable than chlorophyll, the green of herbage. * * * Now not only does chlorophyll abound in many ambiguous microscopical organisms of fresh and salt water, which except for this would be taken for animals, but it has recently been detected in hydras and sea-anemones and planarias, which are as certainly animals as are oysters and clams."

Again Prof. Asa Gray remarks: "The characteristic features of an animal were mouth and stomach. * * * But nature, with all her fondness for patterns, will not be arbitrarily held to them. Entozoa feed—like rhizophytes; and turbellarias and their relatives have no

---

12. Natural Science and Religion, pp. 10-19.

alimentary canal,—the food taken by what answers to mouth passing as directly into the general tissue as does the material which a parasite root imbibes from its host or an ordinary root from its soil." Moreover, Prof. Gray remarks : "The rule is that vegetables create organic matter, and animals consume it, producing none. But, while some animals produce some organic matter, some plants * * * feed wholly upon other plants, or even upon animals or their products." * * * "Again Prof. Gray remarks : "Not many years ago it was taught that plants and animals were composed of different materials : plants, of a chemical substance of three elements, carbon, hydrogen, and oxygen ; animals of one of four elements, nitrogen, being added to the other three. * * * But it was soon ascertained that this quarternary matter of the animal body was chemically the same in the plant, was elaborated there, and only appropriated by the animal. Next it was found that it was physiologically and substantially the same in the plant, that it was *the* living part of the plant, that which manifested the life and did the work in vegetable as well as in animal organism. This substance * * * has, in its state of living matter, one physiological name which has become familiar, that of *protoplasm.*

In view of these facts, Prof. Gray concludes : "The best, I am disposed to say the settled, opinion now is, that there are multitudinous forms which are not sufficiently differentiated to be distinctly either plant or animal, while, as respects ordinary plants and animals, the difficulty of laying down a definition has become far

greater than ever before. In short, the animal and
vegetable lines, diverging widely above, join below in a
loop * * * The fact is, that a new article has recently
been added to the scientific creed, the essential oneness
of the two kingdoms of organic nature."

Here Prof. Gray expresses the opinion of many modern
naturalists who try to explain all mysteries of the organic nature with the help of the microscope, or chemical
analysis.

Yet, these gentlemen forget that the mysterious causes of natural phenomena, which we call *forces*, cannot
be reached with the hand, the eye, — or any, no matter
how ingeniously constructed, instruments. Moreover,
they forget that it is not the material composition of
the organisms, but just those mysteriously hidden *forces*,
that cause all the differences of vital phenomena in
the organic world. These naturalists are consequently
on the wrong track, — and their instruments and observations will only help to make confusion worse—if they
want to find, in the material compositions, the essential
differences between plants and animals.

Many centuries ago, Aristotle and the Christian philosophers, as St. Thomas and his followers, have seen
more clearly into this subject. Their doctrine on this
point is briefly thus explained by the Jesuit Father
Liberatore :[13] "The essential difference by which the
animal kingdom is distinguished from the vegetable, is
sensation (sensibilitas)." Now here our naturalists have
a sure criterion : every organic being which is endowed

---

13. Institutiones Philosophicae, Rome, 1861, pp. 521-527.

with the faculty of sensation,—but not with the power only of making, under certain conditions, some movements, without perception of any kind, as for instance, the Mimosa Pudica (Sensitive Plant) and other plants, —is an animal. Every organic being which only grows and propagates its kind is only a plant.

In order to find out whether any organic being is a plant or an animal, our natualists need not trouble themselves about its material composition, but about the fact whether it has any real sensation, or feeling, or not.

Emil Du Bois-Reymond remarks in the famous address referred to before: " With first feeling of pleasure or pain, which, at the beginning of animal life on Earth, a most simple organized being felt, an insurmountable chasm has been made." Sensation, then, is the chasm that separates plants from animals. — There may be organic beings concerning which reasonable doubts may exist, whether they are animals or plants ;— but only so long as there is any doubt whether they have real sensation, or not. Moreover, there may be animals that look a good deal like plants ;— but let it be born in mind that even the most perfect animals are also plants,—that is, they have also a vegetative life ; like plants, they grow from a seed, develope themselves, and by seeds, propagate their kind.

Having considered wherein the difference between plants and animals consists, let us next see wherein man differs from the animals. As the animals possess all the essential vital properties of plants, so again man possesses all the perfections of the most perfect animals.

But man possesses also a faculty not found in animals—reason. Philosophers have therefore defined man to be an "animal rationale" — a rational animal. Since this question is now-a-days of great importance, it deserves especial attention.

VI. PSYCHOLOGY.

1. ANIMAL SENSATION AND HUMAN INTELLIGENCE.

SCIENTISTS inclined to materialism endeavor, if that were possible, to explain away, not only the barrier that separates the vegetable from the animal kingdom, but also the chasm that exists between Animal Sensation and Human Intelligence.

If, as Darwin, Spencer, Huxley, Haeckel, and numerous others, assume, man were but a highly developed animal, then there would, of course, be no essential, but only a gradual, difference between the sensations of a dog or a monkey, for instance, — and the intellectual perceptions of man. — This false view is held by many modern scientists, and, more or less openly, published in countless works, pamphlets, and periodical publications.

To get at the exact truth on this point, it is necessary to distinguish carefully between what man has in common with the, especially more perfect, animals, and the faculty with which he alone is endowed. Then it will be seen that man is as distinct from a mere animal, as an animal is distinct from a mere plant.

What, then, has man in common with the more perfect
animals ? — The faculty of sensation with its various
ramifications. This faculty is based, as St. Thomas observes, on the sense of touch or feeling, "Sensus tactus
* * * quasi fundamentum aliorum sensuum." [1] Even the
lowest animals, that have no eyes, ears, etc., are endowed
with this faculty, which is admirably perfected in the
higher animals, that have eyes, ears, etc., through which
they receive distinct impressions from material objects.
Moreover, the organisation of every animal is so constructed by the Creator, that to every such impression
of which animals become aware, a certain re-action in
the animal system corresponds. This re-action implies what Christian philosophers call "appetitus sensi-
tivus,"—the sensitive appetite or inclination. This may
be either positive, as, for instance, when a thirsty animal
sees water for drinking, — or negative, as, for instance,
when a hen sees a hawk, and endeavors to escape
the danger. — Something to some extent similar
we may observe with a magnetic needle, the
poles of which are either attracted by some objects, or
repelled from them. Something similar may be seen
with plants, which try to grow towards light, and to get
out of darkness. — But neither the magnetic needle nor
the plants *feel*, as the animal does, this positive or negative inclination.

The more perfect animals, as dogs, elephants, etc.,
have not only the exterior senses of sight, hearing,
taste, smell, and touch, but they are also endowed with

---

1. Liberatore, Institutiones Philosophicae, p, 569.

the power of perceiving, not only certain qualities of exterior objects, but also the dispositions and immutations of their own bodies.

This interior sensation has been called by St. Thomas Aquinas *"sensus communis,"* — and is considered to be the centre in which all exterior sensations meet.[2] This so-called "sensus communis" in animals seems to correspondent somewhat with what in rational and reflecting man is called consciousness. Now, this "sensus communis" not only receives the exterior sensations, but also retains them, at least for some time; — and here we have what we call the sensitive *memory* of animals. Moreover, since animals thus retain the impressions caused by exterior sensations, called *phantasmata* by Christian philosophers, and since certain sensations recall in the memory corresponding former ones, — animals have also what we call *imagination*, or, as Christian philosophers often call it, *phantasy.* — That the more perfect animals are endowed with this faculty, no one will doubt, who ever noticed a dog dreaming, or saw a cat carefully watching a hole into which, as she scents, a rat must have crawled, *although she never saw that rat.*

Finally animals are endowed with a sensitive faculty, called by the Scholastics *vis aestimativa,* which enables them, independently of the sensation of the agreeable, to perceive instinctively what is convenient or repugnant to their peculiar nature.

---

2. Liberatore: Institutiones Philosophicae, p. 566.

These are all the essential psychological faculties with which the most perfect animals are endowed ; — and all these faculties man has in common with them.

But besides these faculties based on sensation, man is endowed with another, a higher faculty, which no animal possesses, — namely, the intellect, or, as this faculty is also called, reason.

This faculty has been termed very appropriately—"intellectus"—from "intus," or "inter," and "legere,"—"to read within,"—or "between." For man not only perceives sensations, as the animals do, but he has also the faculty of penetrating beyond these sensations, and of reading many things in the phenomena presented by the senses, which the senses cannot perceive.

Dr. Paul Haffner[3] explains the doctrine of Christian philosophy on this point as follows : The first and fundamental activity of the intellect we call abstraction, and the thereby received abstract representations — general ideas (universalia) * * * The abstractive activity of the intellect developes itself in three degrees. We abstract from the representations of the senses, first the realities peculiar to every one of the senses. Whilst the nerve of the eye, in consequence of the impressions caused by some rays of light, produces some sensations of sight, the intellect forms the idea of color, light, and figure ; whilst the nerve of hearing perceives the impressions caused by vibrations of the air, the intellect rises to the general idea of sound. * * * Above these abstractions which transform the representations of the various senses into general ideas, arises a second degree

---

3. Grundlinien der Aufgabe der Philosophie, 1881, pp. 193-6.

of abstraction, by which material objects are comprehended in such ideas as are common to all representations of the senses. We comprehend them in mathematical ideas of size, number, motion, etc. Going another step higher, the human intellect forms the antological ideas, by which we comprehend all things in their common essence, — and by their analogies we elevate ourselves to the knowledge of supersensual things and of God. By means of these general ideas, in which the essence of things is comprehended in a supersensual manner, the human intellect is, moreover, enabled to obtain the data by means of which it comprehends the relations of the various objects of existence."

In this manner, then, our intellect reaches, as Liberatore[4] observes, by its own activity, quite immaterial objects, as, for instance, God, the spirits, justice, science, and numberless other objects, quite beyond the province of matter.

Of all this we find no trace in any animal. Therefore Christian Philosophy has always held that man alone in the visible nature has a faculty called the intellect, by which he is elevated as much above the animal, as the animal is above the plant.

It may be mentioned here, that the *intellect* is the source of the following faculties or mental phenomena: First, *reason;* which, as Libertore[5] remarks, is no faculty different from the intellect, but the intellect itself, performing a certain function, which we call reasoning. Secondly, *reflexion;* for the intellect can

---

4. Institutiones Philosophical, Roma, 1861, p. 580.

5. L. c., p. 592.

consider its own acts ; what senses cannot do so. The eye, for instance, can see, but it cannot see that it does see. Thirdly, the *intellective memory ;* for the intellect not only forms ideas, but also preserves them in its treasury. Fourthly, *attention ;* for the intellect can freely turn its acts of thought to certain objects, no matter how distant or non-material. Fifthly, *association of ideas ;* for the intellect can not only reflect on certain ideas, but also, at the same time, recall other ideas, connected with the former, which had been stored in memory.—A faculty of the human mind, specifically distinct from the intellect, yet subordinate to it, is the will ; for we can not only discern what is desirable or not, but also freely choose between different objects.

These are the spiritual faculties of man, which may be improved by using them. Liberatore says : " Experience teaches, that by often repeating certain acts (of thinking, willing, etc.), we acquire a certain facility of acting in that direction : and this facility of acting we call habit." A habit, as St. Thomas[6] teaches, is midway between mere potency and pure act. (Habitus quodammodo est medium inter potentiam et purum actum.)

Having explained the theoretical views of Christian Philosophy on Animal Sensations and Human Intelligence, we will next show that these views are based on carefully observed facts.

----

6. Summa Theol., I. P., Quaest. 79, Art. 7.

2. ANIMAL INSTINCT AND HUMAN REASON.

We have seen in the preceding article, that man possesses all the sensitive faculties with which the most perfect animals are endowed. On this point all agree. Moreover, man possesses a higher faculty which, in the scale of created perfection, elevates him as far above the most perfect animal, as this, by its faculty of sensation, is elevated above the vegetable kingdom. — This faculty, with which man alone in the visible nature is endowed, is called intelligence,—or reason.

As explained before, with this faculty man transcends the realm of sensations, perceives general truths, or abstract ideas,—and penetrates to the very Source of all truth — God, the Creator, Himself. — Nothing like this faculty, Christian Philosophy holds, is to be found with animals ; they possess no faculty beyond the sensitive "actus compositi,"—as Scholastics call them ; — that is, the acts exercised by means of corporal organs. Animals are not capable of perceiving abstract truths, or general principles ; hence they also do not act under the guidance of *abstract*, or *general*, ideas, — but their highest psychological operations are caused by, and confined to, *concrete*, or *individual* objects of reality or imagination, — as careful and exact observations prove.

Since there exists a great confusion of ideas on this point among English and American scientists who confound sensation with intelligence, and instinct with reason, — and consequently see no essential, but only a gradual difference between man and animals, let us examine this subject more in detail.

To avoid confusion, let us *first* consider such psychological operations as are common to all animals of the same species or variety; and *next*, such as may be observed with individual animals, in peculiar circumstances. We will then see that there are no proofs to be found which would warrant the assumption that any animal has a faculty transcending the sphere of mere sensations, and penetrating into the realm of general principles, or abstract ideas.

First, then, let us turn our attention to the so-called instincts which are common to all animals of the same species or variety.

Dr. Ludwig Schuetz[7] mentions the following facts: All animals of the same species or variety select the same kinds of *places for their abodes:* the lion, the desert; the chamois, the summits of mountains; some species of birds inhabit dense forests; — other varieties prefer mountainous regions, etc.— Some species of animals, as the bee and the beaver, *build their own abodes;* and all individuals of the same species build exactly alike. Young birds of the most different kinds are annually hatched; they never saw that, or how, their nests were built. But when spring comes again, they go to work, select a place, collect materials, and build a nest, in all respects just like the one in which they first saw daylight. — Every species of animal has its own peculiar kind of *food:* swallows, insects; squirrels, nuts; cattle, grass and grain; etc. If some animals get sick, they change their diet; cats and dogs, contrary to their ordinary appetite, in such cases, eat grass.—Some animals

7. Das Thier hat keine Vernunft. Muenster, 1871.

have to get their peculiar food at a great distance: the green-peak of North America, at the proper season, crosses the Atlantic Ocean, to get his share of the French cherries; other birds cross the great Sahara, the Mediteranean Sea, etc., in quest of food; and at the right time they return home again. — Some animals that live on food which cannot be found in winter, collect a sufficient supply beforehand, which they, as it would seem, wisely and intelligently store away for winter's use; as, for instance, the squirrel, the hamster, the whistling rabbit of Siberia, etc.—Some animals develop a great deal of ingenuity in capturing, or otherwise procuring their food; for instance, some ants and the squirting fish of India; lobsters slyly watch oysters, and let little pebbles drop between the shells, as often as these are opened, until the oysters can be got at, etc. — *In case of danger*, all animals of the same species protect or defend themselves and their young ones in a like manner. Rabbits, deer, and other timid animals flee; the partridge, young chickens, etc., hide in the grass, or cower as much as possible, so as not to be seen.—Some animals are provided with weapons of defense: the bee, with a sting; the eagle, with a beak and claws; "bulls aim there horns, and asses lift their heels,"— as a poet remarks, etc. — Most remarkable are the instincts of the various species of animals in regard to the *propagation of their kind.* — The young of each species are brought forth at a time most favorable to their growth. Many species of animals prepare convenient *resting places* for their young: birds build various kinds of nests; the rabbit collects fine moss, etc., for that purpose; the turtle and crocodile lay their eggs in warm sand near water; many insects deposit their eggs under leaves which protect them from birds and provide their young, as soon as hatched by the warmth of the atmosphere, with food. — Many animals exhibit a marvelous sagacity in feeding their young. — Some species of birds are very careful not to betray their nests. Before flying to it, they first stop for a while at a good distance from it, and watch whether they are being observed, or not. Just as circumspect they are, when again leaving the nest. If there are several young ones to be fed, the parent-animals observe a strict order, so that no one is fed more than once at a time, and no one overlooked.—Some of the older birds, as sparrows, live mostly on seeds; yet as these would be too hard to digest for the young ones, these are fed, at first, always with tender insects.—Still more remarkable it is, how carefully many insects that will not live any more by the time their young come to life, provide for these beforehand, so that they will find right off the food they need. —When *danger threatens their young,* some animals show a skill in defending them, that would be creditable to any rational being under the same circumstances. For instance, when cats, dogs, owls, etc., perceive that their young are endangered in a certain place, they will carry them to a more secure place. In such a case, the young of some species get willingly on the backs of the old ones, and let themselves be carried away. The hen will sound an alarm, and the little chickens taking the hint, will hide as quickly as they can under the wings of the old hen, or in high grass, etc.—A male partridge, when a dog or a man comes near its nest, will by its cry warn the female and the young; then act as if it could not fly away, to attract the enemy's attention, until the female and the young are in safety.—Not less remarkable are the methods by which some species of animal protect themselves against surprise. For instance, the chamois, flamingoes, etc., wherever they stop, have regular sentinels stationed about, that, in case of danger, give a warning signal, at which all follow the the leader of the flock, or act otherwise in accordance with their peculiar instincts.

Numerous such illustrations could be added, which show beyond a shadow of doubt, that every species of animal, in many respects, acts according to principles of prudence,—principles of reason.

Now the question is: Does this prove that animals are endowed with intelligence,—with reason?

This is affirmed by numerous modern writers, especially such as endeavor to explain away the essential differences between animals and man. These writers conclude:—because animals act according to principles of reason, to attain certain desirable ends, therefore they must be endowed with reason.

The fallacy of this conclusion will be easily perceived. Machines, too, operate according to principles of reason, to attain certain desirable ends;—but who would therefore assert that they are endowed with reason?—Moreover, in the vegetable kingdom we find everywhere that plants operate according to principles of reason, for certain desirable ends.—Plants need food; the roots of plants invariably endeavor to extend themselves in the direction *where* the most suitable plant-food is to be found.—Plants need light; if they have not got their fair share of it, they do their best to get as near the light as possible; as plants growing in partially dark cellars, under the shade of trees, etc.—Some plants even catch animals and use them for food. Prof. Asa Gray [8] relates of the plant *Dionaea*, that its two sides suddenly close like a trap, to enclose an alighted fly; then it pours out digestive juices, and in due time re-absorbs the whole.—Some plants, as grape-vines, tend to grow upwards, as high as possible. Their own stem is too weak to support them. Hence they stretch out miniature fingers in every direction—the tendrils, to get hold of something, to lift themselves up higher. As soon as a tendril has reached a convenient hold, it will commence to twist itself firmly around it.

Now, such and numerous similar illustrations which could be added, show beyond doubt, that also plants, like animals, act according to principles of reason—to attain certain desirable ends. But who would therefore dream of ascribing reason to plants?

In fact, wherever we look to in the wide universe, we see all beings operating according to principles of reason—for certain ends. The Sun, the planets, etc., move according to exact laws; the smallest atoms do the same. Every plant grows and brings forth fruit of its kind without fail. Every animal lives, grows, moves, and acts, within its sphere, with the regularity of a delicately constructed clockwork or machine.

8. Natural Science and Religion, p. 23.

This, no doubt, proves that there is *a rational cause* of these operations,—but not that either the plants or animals have reason.—That plants have intelligence, or reason, probably nobody of sane mind will assert; that animals, notwithstanding their seemingly quite rational operations, have no intelligence, or reason, but are actuated by an irresistible, unconscious impulse,—called instinct,—will become plain to every one who reflects on the following facts to which Prof. Joseph Le Conte[9] calls attention.

After describing some most remarkable phenomena of insect life, he continues: "Now, such actions performed by man would show high intelligence and much experience; and yet we cannot attribute such intelligence to these insects, because their actions in other directions and under other and new conditions exhibit but a very small amount of intelligence (?); we are compelled to attribute these wise actions to another and somewhat different faculty, which by way of distinction we call instinct." "Intelligence," he adds, "works by experience, and is wholly dependent on individual experience, for the wisdom of its actions. * * * On the contrary, instinct is wholly independent of individual experience. The young bee or mud-wasp, untaught, works at once without hesitation, with the greatest precision and in the wisest manner, to accomplish the most marvelous results."

Morever, "*Intelligence belongs to the individual, and is therefore variable, i. e., different in different individuals, and also improvable in the life of the individual by*

9. The Popular Science Monthly, October, 1875, p. 657.

experience.—Instinct *belongs to the species*, and is therefore the same in all individuals and unimprovable with age and experience. * * * *Instinct in its sphere is far more perfect and unerring than intelligence.* — It makes no mistakes, because determined by structure, not by imperfect knowledge.—In a word, intelligent conduct is *self-determined* and becomes wise by individual experience. Instinctive conduct is *predetermined in wisdom by brain-structure.*"

This explanation, for what is true of insects in this regard, is true of animals generally, suffices to show the vast difference between Animal Instinct and Human Intelligence, or Reason. Human Reason is a faculty above the sphere of sensation ;—a faculty which looks down upon sensations, as the eye looks upon visible objects. Man is not only aware, like animals, of external and internal sensations, but by his intelligence, he also penetrates into the nature and objects of these sensations ;—which no animal can do.

Human Intelligence, or Reason, therefore, though it operates on facts of sensation, is not bound to corporeal organs, but is a free, self-determining faculty; whereas Animal Instinct is most intimately connected with the animal organism, therefore entirely dependent on it in its operations. Like sensation, Animal Instinct depends on two factors : a sensitive impression on the organism,—and a reaction of the latter. Human Intelligence, or Reason, is a free, spiritual faculty ;— Animal Instinct, on the other hand, a pre-determined merely sensitive faculty.

St. Thomas Aquinas [10] remarks: "No operation of the sensitive part (faculties) can take place without the body (organism). In the souls of animals, no operation beyond those of the sensitive part are to be found; for they (animals) neither understand nor reason; which is evident from the fact that all animals of the same species operate alike; for all swallows build their nests alike, etc. Therefore, there exists no operation of the animal soul, which could be accomplished without the body (organism)."

Against this doctrine of Christian Philosophy,—that animals possess only sensitive faculties, but no real intelligence or reason, some modern scientists advance the following objections:

They claim to have detected slight differences in the skill of different bees of the same varieties;—also, that some kinds of birds built their nests slightly differ-ent from what they did formerly.

Let us suppose that such discoveries are based on facts existing outside of the imagination of their dis-coverers,—what do they prove? Not more than that, owing to some unknown influences, a slight change has been effected in the organisms of those animals,—and this change has called forth also a change in their way of operating.

But that is nothing new.—Still greater differences in instinct may be observed among other animals; for instance, among dogs of the same breed.—But such differences do not prove that at least a "small margin

---

10. Summa Philosophica, Liber. II., Cap. 82.

of intelligence" exists in these animals ;—all what they prove is, that the organism of such animals has been slightly changed, consequently also their instincts and modes of operation.

Another objection often raised, is the following : By training, some animals, as dogs, horses, elephants, parrots, monkeys, canaries, and even—fleas, can be taught various amusing performances.—This proves, some imagine, that these animals possess intelligence. We answer : Intelligence has nothing to do with the learning of such performances ; — the sensitive memory it is, to which persons training such animals can, and do, appeal.—Mr. Rarey, the famous horse-tamer, who knew more about the real nature of a horse's so-called mental faculties, than many a theoretic scientist, was in the habit of prefacing his performances by a short lecture, in which he used to say—that he conquered the animals, "because he possessed reason, and the horse did not." [11]   What is true of the horse, is true of every other trained animal.

Some scientists imagine to have discovered actions of animals, which cannot be explained without admitting that animals have intelligence.   In an article on " Intelligence in Animals," R. A. Proctor [12] relates several cases to the point ; for instance : monkeys were exceedingly careful in handling sharp tools, after having been cut once ; or, in opening paper-bags, after having been stung by a wasp concealed therein ; the muleteers in South America say, " I will not give you the mule whose

11. Man and Beast, by Rev. I. G. Wood, New York, 1875, chapt. 2.
12. Nature Studies, New York, June 18, 1883, p. 118, etc.

step is easiest,—but *la mas racional* (the most rational) ; " a rat, after noticing birds continually fluttering about a window-sill, where they were fed, managed, with the help of a shrub, to climb up there, although it was a height of 13 feet ; some rats, to get at pure water, gnawed a hole in a leaden pipe ; a cat made use of the knocker of a side door, whenever it desired admission ; a small English terrier which had been taught to ring the bell for the servant to come, would not ring when the servant was in the room ; etc.

Such and similar instances are often appealed to, to prove that animals have—intelligence. Yet, there is nothing in such cases, that cannot be explained in accordance with the doctrine of Christian Philosophy, that no animal operation goes beyond the *sensitive* faculties. Even some modern scientists are gradually coming to this conviction. R. A. Proctor remarks, the theory has been " recently advocated by Huxley and others, that animals are automata, not possessing consciousness ; "—by which, perhaps, substantially the same faculty is meant, which Christian philosophers would call intelligent reflection.

In the article referred to before, Mr. Proctor remarks : " Mr. Henslow makes a good point in noting how like the practical reasoning of animals, is the reasoning of young folks. ' A boy the other day,' he says, ' found the straps of his skates frozen. The fact only suggested *cutting* them. Not one of his school-fellows reflected upon the abstract fact that the ice would melt, if he sat upon his foot a few minutes. Hence brutes and boys are exactly alike in that nothing occurs to either beyond what the immediate fact before them may suggest. The one kind I call purely *practical* reasoning, which both have ; the other abstract, which brutes can *never* acquire ; but the boy *will*, as his intelligence develops."

Our modern scientists may be somewhat surprised to learn that substantially the same view has been held by Christian philosophers since many centuries. * * * St. Gregory of Nyssa, who lived about fifteen centuries ago, as quoted by St. Thomas Aquinas,[13] observes, that children and irrational beings (brutes) act voluntarily but not with free (intelligent) choice;"—that is, children whose intellectual faculties are not yet developed, and brutes, act in consequence of impressions on their sensative faculties, and not from free, abstractive intelligence, or reason.

We therefore conclude, that Animal Instincts, or the operations of brutes, never go beyond the sphere of the sensitive faculties, whose exercise is completely dependent on the organism, — and that between Animal Instinct and Human Reason there exists a similar *essential*, not only gradual difference, as between the vegetative life of plants, and the sensitive faculties of brutes.

Some modern scientists but betray their ignorance, when they make assertions like the following :[14] "The intelligence of animals, if not equal to man's, is at least like it, and * * * the difference between the oyster anchored to its rock and the *homo sapiens* of Linnaeus are merely differences between more or less, degrees of succession that make up what is called the scale of being.

13. Summa Theol. Prima Secundae, XIII, II.
14. See The Popular Science Monthly, April, 1874, p. 729.

It is the latter opinion that has been declared (by whom?) triumphant (?) by the researches of natural history and those of comparative anatomy alike."

Such unfounded and false assertions are often made by some writers on what they call "animal intelligence." Their false views on this subject have their origin in the confusion of the merely sensitive with the strictly intellectual faculties. Man possesses both kinds of faculties ; — being in "the scale of being" intermediate between pure spirits, or angels, and brutes, he partakes of the nature and faculties of both. Animals possess the sensitive faculties alone : the five external senses, — the sensus communis, as the Scholastics term it, which we may call, the centre in which the external sensations meet,—the imagination,—*sensitive* memory,— and the sensual appetites and instincts.— All these faculties are dependent for their operations on the corporal organism.—What animals do not possess, is intelligence, or reason, — free or intelligent will, — and intellectual memory.

The best advice that can be given to such modern scientists as wish to be cured of their confusion of ideas, concerning the psychological faculties of man and animals, is—to study the works of St. Thomas Aquinas, or of other sound Christian philosophers on this subject.

A few remarks on the so-called "aesthetic sense in animals" may be not out of place in this article. Some modern scientists, to prove that animals have intelligence, call attention to what they term "the aesthetic sense in animals." — "In their view," a writer [15] re-

_____
15. The Popular Science Monthly, April 1874, pp. 729, etc.

marks, "just as animals are endowed with intelligence as well as man, though in a lower degree, so in the same way and in the same proportion they are endowed with sense of beauty. They find the proof of this rather bold assertion * * * in sexual selection * * * Among all animals, they say, among insects, fish, birds, mammals, the male chooses his female, and the female chooses her male. If strength often determines the choice, so beauty often does too. The charms of graceful shapes, pleasing colors, fine notes, has great weight in settling the preference. * * * Hence all those displays of genuine coquetry, which may be easily observed, in pairing time, among all animals; hence those attractions prevailing through vigor of form, brilliant hues, and impassioned song. This general fact * * * gives * * * ground for asserting that animals, having the perception of beauty, have consequently the aesthetic sense."

If our modern scientists will keep clearly in view, as explained before, the difference between sensitive impressions and intellectual perceptions, they will easily see the vast difference between the sensual emotions caused in animals, and the intellectual pleasures called forth in man by the beauty of objects. — Animals through their senses perceive only beautiful objects; but man sees not only beautiful objects, but by an act of intellectual judgment he perceives also the beautiful *in abstracto* in beautiful objects; — which no animal does perceive. — Some beautiful objects, — individuals of the opposite sex of the same kind —, no doubt, do make a certain impression on the imagination of animals, and *through the imagination on the sensual appetite.* But this impression does not by any means go beyond the sphere of sensation. The very fact that the only imaginable object of this impression is — to bring individual animals of the same species together, — and usually only at a certain time of the year, — shows that this impression is something *merely sensual;* — whereas man by his superior intellectual faculty perceives the beautiful in objects, not only in individuals of his own kind, but wherever it exists in nature ; not only at a certain time of the year, but always.

Thus we see, that in man the aesthetic faculty is not dependent on certain sensitive influences alone, exercised on the organic structure, and that it is always active, wherever and whenever anything beautiful appears to him.

How different it is with animals. Mr. Charles Leveque says : [16] Place your animal before a work of art, representing its male or its female, with a precision that deceives the eye * * * A dog would perhaps stop a moment in front of Oudry's hunting-pieces, if their frames were put on the floor, within reach of his look. He would come up, examine them, ask the canvass a single question with his infallible *scent* — and that would be all * * * It is not the expression of life in general that he wants ; it is life itself, individual life, life which speaks to his senses, and to that of smell much more strongly than to his eyes and ears. He has no concern with the general, the ideal, the admirable ; he understands nothing about them."

16. The Popular Science Monthly, April, 1874, p. 731.

We therefore conclude that the so-called "aesthetic sense in animals" is no more than the result of merely *sensual* impressions caused by seeing, etc., some individual of the same kind ; and the end for which the Creator has endowed animals with this sense, is no other than to direct the sexual selections of animals according to certain rules. A higher end, or object, this so-called "aesthetic sense in animals" has not. Quite different it is with the aesthetic faculty of man ; its objects are by no means, as with animals, only individuals of the same species and the opposite sex, — but *all* perceptible things beautiful in the physical, moral, and supernatural order. Man not only sees, like animals, beautiful objects, —but by his intelligence he perceives also the beautiful— *in abstracto* — in the objects. The end for which the Creator has endowed man with the aesthetic faculty is not merely, as with animals, to direct sexual selection, but to raise him by the contemplation of created beauty to the source of all beauty, the Creator ; the vision of Whom shall for all eternity constitute the happiness of man.

It is natural, then, for men to admire and enjoy what is beautiful in creation ; but they should always keep these words of the Book of Wisdom (13, 3) in mind : "Let them know how much the Lord of them is more beautiful than they (the created beings) : for the first Author of beauty made all those things."

3. THE RELATIONS OF BODY AND MIND IN MAN.

Having examined the *essential* differences existing between man and brutes, let us next turn our attention to the relations of body and mind in man.

As the reader will remember, Christian Philosophy holds that in all visible beings there is a passive and an active principle,—which the Scholastics have termed "materia" and "forma." —In man, the body is the passive principle,—the mind, or soul, the active. Yet both are so intimately united as to constitute but one complete being, man.—Plato assumed that the relations between the soul, or mind, and the body, were somewhat like those existing between a rider and the horse on which he rides.—But here we have two distinct *complete* beings, the horse and the rider. Christian Philosophy teaches, that body and soul constitute but *one complete* being,—man. Nevertheless, to some extent, Plato's comparison of the soul and body—to a rider and his horse, may be admitted.—As a rider exercises some influence over the horse,—and the horse on the rider,—so does the soul over the body, and the body on the soul, though their mutual influences are incomparably more subtile and mysterious than those of the former.

Alexander Bain, Professor in the University of Aberdeen, Scotland, says[1] on this subject: "The facts showing that the connexion of Mind and Body is not occasional or partial, but thorough-going and complete, are such as the following: In the first place, it has been noted in all ages and countries, that the Feelings

1. Mind and Body—The Theories of Their Relations.

possess a natural language or Expression. * * * On this uniformity of connexion between feelings and their bodily expression depends our knowledge of each other's mind and character. * * * To the painter, the sculptor, and the poet, every feeling has its manifestation. * * *

"A second class of proofs of the intimate connexion between Mind and Body is furnished by the effects of bodily changes on mental states, and of mental changes on bodily states. * * * As to the influence of bodily changes on mental states, we have such facts as — the dependence of our feelings and moods — upon hunger, repletion, the state of the stomach, fatigue and rest, pure and impure air, cold and warmth, stimulants and drugs, bodily injuries, sleep, advancing years. These influences extend * * * to the highest emotions of the mind—love, anger, aesthetic feeling, and moral sensibility. * * * Bodily affliction is often the cause of a change in the moral nature.—The bodily routine of our daily life is the counterpart of the mental routine. A healthy man wakens in the morning with a flush of spirits and energy. * * * Toward the end of day lassitude sets in, and fades into the deep unconsciousness of healthy sleep.

"Since the intellectual faculties appear to be most removed from the effect of physical agencies, I will quote a few facts, showing that in reality they have no exemption from the general rule.—The memory rises and falls with the bodily condition; being vigorous in our fresh moments, and feeble when we are fatigued and exhausted. Old age notoriously impairs the memory in ninety-nine men out of a hundred. * * * Why should sleep suspend all thought, except the incoherency of dreaming (absent in perfect sleep), if a certain condition of the bodily powers were not indispensible to the intellectual functions?

"The influence of mental changes upon the Body is supported by an equal force of testimony. Sudden outbursts of emotion derange the bodily functions. Fear, paralysis, digestion, great mental depression, enfeebles all the organs. Protracted and severe mental labor brings on disease of the bodily organs. On the other hand, happy outward circumstances are favorable to health and longevity.

"In considering minutely the evidences of the connexion of mind and body, we gradually perceive that the organ most intimately associated with the mind is the Brain * * * Yet, although the Brain is by pre-eminence the mental organ, other organs co-operate; more especially the Senses, the Muscles, and the great Viscera * * * (The Brain) is a very large and complicated organ; it receives a copious supply of blood, computed as one fifth of the entire circulation * * * Now the facts that connect the mind with the brain are numerous and irresistible. Let us rehearse a few of them, under the two aspects; brain changes affecting the mind, mental changes affecting the brain.

"Under the first topic, the commonest observation is the effect of a blow on the head, which suspends for the time consciousness and thought; at a certain pitch of severity it produces a permanent injury of the faculties, impairing the memory, or occasioning some form of

mental derangement. It may also *remedy* derangement ;
there are cases on record, where a blow on the head has
cured Idiocy.

"All those abuses and casualties that impair the mental
faculties act upon the nervous substance. Thus, stimu-
lating drugs operate upon the nerves. Many instances
of imbecility of mind are distinctly traced to causes affecting the nutrition of the brain. The more careful and
studied observations of physiologists have shown beyond
question that the brain as a whole is indispensible to
thought, to feeling, and to volition ; while they have
further discriminated the functions of its different parts.

"Next, as regards mental changes leading to brain
changes, or being associated with them, we can quote
very extensive observations. Thus, after great mental
exertion or excitement, there is an increase of the products of nervous waste * * * Again, violent emotions
are among the causes of paralysis, which is a disease of
the nerves or nerve centres. Most decisive of all, under
this head, is the wide experience of the insane. Among
the chief causes of insanity must be reckoned excessive
drafts on the mind — as, for example, long and severe
mental exertion, and sudden mental shocks, usually of
disaster and misfortune, but occasionally even of joy.
The association of brain derangement is all but a perfectly established induction. In the great mass of insane patients the alteration of the brain is visible and
pronounced * * * It is believed, however, that in *all*
cases of pronounced mental aberration, disease of the
brain is present in a marked form.

"A very instructive class of facts may be adduced, connecting mental action with the quantity and quality of the *blood* supplied to the brain. No organ is active without blood. The demand made by the brain corresponds with the extent and energy of its functions. Deficiency in the circulation is accompanied with feeble manifestations of mind. In sleep, there is a diminution of the supply of arterial blood to the brain. General depletion lowers all the functions generally, mind included. On the other hand, when the cerebral circulation is quickened, the feelings are roused, the thoughts are more rapid, the volitions more vehement; great mental excitement is always accompanied with an unusual flow of blood, often outwardly shown by the throbbing of the vessels. In delirium, the circulation attains an extraordinary pitch.

"The blood must possess a certain *quality*, involving the presence of certain ingredients and the absence of others. Wholesome nourishment supplies the first condition of nervous and mental activity; inanition or starvation, feebleness of digestion, militate against the exercise of the mental functions. Moreover, the blood may be abundant and rich in nutritive matters, yet the organ of the mind may be unduly depressed by the excessive draft of the other interests of the system, as, for example, the muscles; under great muscular strain, there is very little capablity of mental effort. Again, there are certain substances, known as stimulants, that are considered to supply the blood with an element especially provocative of nervous change; as alcohol, tobacco, tea, opium, etc.

"The substances that must be absent include the so-called poisons, and the impurity of the blood itself, which several large viscera are occupied in removing. The chief of these impurities are carbonic acid and urea; either of them left to accumulate in the blood leads to mental depression, unconsciousness, and finally death. Hence the mental tone depends no less upon the vigorous condition of the purifying organs — lungs, liver, intestines, kidneys, skin — than upon the presence of nutritive material obtained from the food."

These facts, to which many details could be added, are quoted from the well known work of the late Prof. Alexander Bain, of Scotland, who has the reputation of being one of the most prominent modern psychologists, to give the reader a clear idea of the most intimate connection existing between the body, especially the brain, and the mind of man.

Since modern infidel scientists often appeal to such facts, to impugn the Christian doctrine of the immortality of the human soul, it is necessary to understand clearly, — that although the most intimate connexions exist between the body and the mind,—these are yet *per se* two as distinct realities, as the rider and his horse, or as the musician and his instrument.

As explained before,[2] the soul of man is the so-called "forma" of the body; but it must not be forgotten that it is the most perfect "forma" of any visible being. — The great theologian and philosopher St. Thomas Aqui-

1861, pp. 654-5.

---

2. See Institutiones Philosophicae, by M. Liberatore, S. J., Romae,

nas [3] says on this subject. "The more noble a *forma* is, the more it rules over corporeal matter, and the less it is immerged in the latter * * * The higher the nobility of the *formae* is, the more we find that their power exceeds that of elementary matter,—as the vegetative principle (anima) exceeds the elementary *forma*, and the sensitive soul, the vegetative principle. But the soul of man is the most exalted of the *formae* in the scale of perfection. Therefore it exceeds by its power (virtus) corporeal matter to such an extent, as to possess a certain faculty, of which corporeal matter has no part whatever, and this faculty is called the intellect."

From this we see that Christian Philosophy not only teaches the most intimate union of body and soul, or mind, in man, but also clearly points out the essential difference between both. — What is true of the *modus existendi* (mode of existence), is also true of the *modus operandi* (mode of operating). Man, in the scale of creation, is intermediate between the brutes and the angels, partaking of the nature of both. Therefore, also, his peculiar *modus operandi* by his intelligence, is far above the faculties of brutes, yet far below the pure intelligence of angels. In human intelligence, to some extent, animal perception and purely spiritual, angelic intelligence are blended. On the one hand, man, like the angels, perceives ideas — what the brutes do not; but on the other hand, man needs *in the exercise of his intelligence* a faculty which he has in common with the brutes,—his imagination;—what angels do not.

---

3. Summa Theologica, P. I, quæsto 76, art. 1.

And here is the point that perplexes many a modern infidel scientist. Because man, in the exercise of his intelligence, continually makes use of the imagination, with which also animals are endowed, — and because, if this imagination is disordered, also the human intellect cannot properly perform its operations,—therefore some infidel scientists claim that the soul of man is not essentially different from that of the animal,—and will consequently once share the fate of the latter—perish.

Here we must remember what was explained in a former article.—In this visible universe there exist these four great divisions of beings, that rise above each other as distinct planes : — The inorganic material elements, the vegetable kingdom, the animal kingdom, and man. Now, it is a general law of this visible universe, that every higher division, or plane, of existence is, in its operations, dependent on the division, or plane, immediately below it. — Thus plants, in order to grow, depend on a fair supply of inorganic elements — air, light, water, and mineral nourishment.—Animals, again, to exercise the faculties of sensation peculiar to them, depend on the good condition of their vegetative nature. A starving animal will loose its faculties of sensation in proportion to its starvation ; in order to be able to exercise its sensitive faculties well, it is necessary that an animal be sufficiently fed.—Now man, in order to exercise his faculty of intelligence, or reason, is dependent on the healthy state of his sensitive faculties,— especially of his imagination, which presents to him the objects on which his intelligence will operate.

Since this point is of the greatest importance, to get a clear idea of the relations of body, or brain, and mind let us examine it more carefully.

As explained before, man possesses not only all the sensitive faculties of the most perfect animals, but be_sides, and beyond these, intellectual, spiritual faculties— intelligence or reason, free will, and intellectual memory. The intellect is the source or foundation of all spiritual faculties of man. It is the intellect by which he obtains the ideas. This intellect, when drawing conclusions from given premises, is called reason. This intellect which guides the selection of objects in certain cases, is the source of will. This intellect, finally, which not only forms ideas, but also retains them, is the intellectual memory.

Now, Christian Philosophy teaches : It is impossible for our intellect, in the present state of existence, whilst it is united with a body subject to change, to understand (intelligere) anything, except by turning its attention on phantasmata, that is, representations of the interior sense and the imagination.—St. Thomas [4] adds : "This every one can observe with himself ; for if any-one tries to understand anything, he forms to himself some phantas- mata (representations of the imagination),—as it were, examples, in which, we may say, he looks at what he endeavors to understand. And therefore also, if we wish to make anything clear to anyone, we propose to him examples, from which he may form for himself phantasmata, in order to understand." — Thus we

4. Summa Theologica, I. P., Quaest. 84; art. 7.

see that, in the present state of life, the exercise of our
intelligence depends on our faculties of sensation and
imagination, somewhat like the eye that sees, on its vis=
ible object. For the imagination presents the objects,
which the intellect is to understand.

Now, we know that sometimes the eye is hindered from
seeing ; for instance, when a cataract has been formed on
the crystalline lens of the eye;—or the eye may see, but
not correctly, as, for instance, if one looks through red,
blue, etc., spectacles. Although the eye may, in such cases,
not see, or not see correctly, yet the eye itself, or more
properly the retina which receives the impressions, may be
perfectly sound. If the cataract, or the colored spectacles
will be removed, the eye will again see correctly.—In a
similar manner it is with the human intellect. The intellect itself, being a purely spiritual faculty, cannot be
subject to disease ; but the imagination, the medium by
which phantasmata are presented to the intellect, being
a sensitive faculty, may become disordered, and then the
intellect will have no proper objects to contemplate.—
Therefore St. Thomas[5] teaches : "We see that when the
imagination is impeded by an injury done to the organ
*   *   *   one is prevented from understanding even
things of which one had a knowledge before."—In such
cases, not only the intellect but also the will and the
memory may become disordered as to their operations.

This, then, explains satisfactorily all those phenomena
concerning the relations between body and mind, to
which modern infidels often appeal as to proofs against
the spirituality of the human soul.

5. L. C.

Long before so-called Modern Science existed, Christian philosophers had explained how the human soul can be spiritual and immortal, although people may become insane, or lose their memory; although they may be affected by alcoholic drinks, etc.; although the intelligence seems to grow and decay with man; etc.

In all such cases, it is not the human intellect that is directly affected, but only its sensitive medium, the organs of sensation and imagination — which are subject to various disorders caused by physical influences.

4.   THE IMMORTALITY OF THE HUMAN SOUL.

What is matter? What is force? What are the mysterious principles that make plants grow, animals feel, and man think? — These are questions which modern scientists can answer no more, than Aristotle could two thousand years ago.

Modern Science has, no doubt, made immense progress in investigating the various *phenomena* of the visible world, — the outside shells of things, we may say; but the intrinsic natures of the principles of activity and life cannot be reached by the microscope or any other conceivable instrument; they remain as mysterious as ever.

For this reason, Modern Science can teach nothing new concerning to the Immortality of the Human Soul. The *phenomena* resulting from the relations of body and mind have centuries ago been, substantially, as clearly known to Christian philosophers, as they are now to our modern scientists; and centuries ago, Christian philosophers, as St. Thomas Aquinas, have triumphantly refuted such objections against the Immortality of the Soul, as some modern infidel scientists imagine to have now been discovered for the first time.

Among the reasons why Christian Philosophy has always maintained the Immortality of the Human Soul, are the following:[1]

Of all visible beings, man alone has a desire for life without end ; he alone can reflect on immortality, and shudders at the idea of annihilation, as at something contrary to the nature of his being. — Whereas animals, even the most perfect, give no evidence of having any idea, or desire, of an existence without end. Being endowed, as was explained before, only with sensitive faculties that depend entirely on material impressions and changes of the organism, animals perceive but objects existing in the present. Concerning a distant future, no animal troubles itself ; it has no idea of personal immortality ; it feels only *present* pleasure or pain.

Now, why should the Creator have endowed man with the faculties of comprehending, and most decidedly longing after, immortality, if this should not be granted to him ?

Moreover, the Creator has, as explained before, endowed man with the intellectual faculties of perceiving and enjoying the true and the beautiful. As the intellect of man craves for the knowledge of truth, so his will craves for the enjoyment of the beautiful. Now, both

---

1. See Institutiones Philosophicae, by P. M. Liberatore, p. 636, etc, and St. Thomas' Summa Philosophica, Liber. II., 82.

of these faculties are never fully satisfied in this life ; they find no rest, until the rest in Him Who is the source of all truth and all beauty.

Why should the Creator have endowed man with these faculties, if they should never be satisfied ?

Divine Revelation (John 17, 3.) teaches : "This is life everlasting : that they may know Thee, the only true God, and Jesus Christ, Whom Thou hast sent."

God alone Who is the Source of all truth and "the first Author of beauty" (Wisdom 13, 3), can fully satisfy the longings of the human soul after infinite truth and the infinitely beautiful.

Another reason which points to the immortality of man, is the very idea of moral order. — The actions of animals are confined within the limits of the physical nature. Animals perform actions, or omit them, in consequence of certain physical impulses, or influences, over which they have no control. Their admirable instinct teaches them what they must do, and what avoid, or flee. A higher object than to gratify their natural impulses, is unknown to them.

Quite different it is with man. He not only feels what is agreeable or disagreeable to his sensitive appetites, but he also perceives a higher, a more sublime law which is to govern his conduct.—His conscience tells him that certain acts are good, others praiseworthy, others heroic, and some bad, contemptible, and damnable. His conscience warns him against doing evil, and encourages him to do good. His will has liberty either to follow or disobey the dictates of his conscience, which always admonishes him to act according to the principles of morality, of the rational order which God has established for intelligent beings to observe. Now, if man's life were to end with the death of his body, it would then, at the end, be all the same, whether a man had observed the moral order established by the Creator, or whether he had trampled every one of its principles under foot. It would, then, be all the same, whether a man would have been a kind father, a charitable friend of the poor, a virtuous husband, a just person in all respects, or whether he would have been a tyrant or a fiend that continually abused his family, that murdered his wife and children, that cheated, stole, and robbed, that committed all impurities possible, that, with one word, was a scourge to all who unfortunately came in contact with him.

Such an idea as that at the end it would be all the same, whether a man had been all his lifetime just, kind, and pure-minded,—or whether he had been a treacherous, cruel, and filthy demon, — is most revolting to the sublime sense of justice, which the Creator Himself has implanted in the soul of every man. If man were mortal, the Creator would, what is impossible, have made Himself guilty of deceiving man by endowing him with this noble sense of justice.

Thus we see that the Creator Himself has deeply implanted in man faculties which point to an existence far beyond the grave,—to immortality. This conviction is so deeply implanted in man, that we find all nations, whether ancient or modern, whether barbarous or civilized, most profoundly penetrated by it.

No doubt, as Holy Scripture (Ps. 13) says: "The fool hath said in his heart: There is no God," consequently no immortality; but let it be remembered,— the fool only "says" so; — he is not convinced of it; — for as the famous philosopher Bacon of Verulam observed:[2] "It appeareth in nothing more, that atheism is rather on the lip than in the heart of man, than by this, that atheists will ever be talking of that their opinion, as if they fainted in it within themselves, and would be glad to be strengthened by the consent of others. * * * If they did truly think that there were no such thing as God, why should they trouble themselves?"

There are, now and then, some unfortunate men who declare that they do not believe in immortality. But their declaration is energetically contradicted by their own conscience, which revolts at the very idea of annihilation.

Abbe Mullois, Chaplain to Emperor Napoleon III., remarks[3] on this point: "It is not rare, indeed, to meet with men who call themselves unbelievers, who assert it, and who write themselves such; but will you find men who are seriously unbelievers, and who do not falter in their negations? A pious priest who was frequently called upon to attend the sick in the higher classes of society in Paris, was once asked whether he often met with men who had ceased to believe. He replied: 'Pray, don't allude to the subject. Though I have been long accustomed to minister great

---

2. Essay on Atheism.
3. The Clergy and the Pulpit, Chapt. 3.

sinners, I have never yet had the good fortune to lay my hand on one who was even a little unbelieving. As regards the faith, men in general are better than their words or their writings either.' — As has been well remarked : The man who, in all sincerity, says : 'I don't believe,' often deceives himself. There is in the depths of his heart a root of faith which never dies."

To this same truth, many centuries ago, the famous ancient writer Tertullian called attention. His observation "Anima humana Christiana"—Man's soul is naturally Christian—is confirmed by the fact that, in spite of a few infidels now and then, all nations preserve religious convictions, even if the expressions of their religious feelings are sometimes wrong.

Therefore we say :—although now and then an infidel may pretend not to believe in immortality, yet this belief has been so firmly imprinted by the Creator in the human soul, that no sophistry is able to destroy it. Although some individuals may, in spite of their revolting consciences, proclaim to believe in their future complete death, yet all nations will continue to cherish firmly .the conviction that the grave is not the final goal of man ; that beyond the grave there is another, a more permanent life, where we are also to meet our dear friends again, that have gone before us.

What now has Modern Science to say on this subject? As remarked before, modern scientists, — with all their boasting about progress, — are to-day as little able, as Aristotle was more than two thousand years ago, to explain what matter or force is, what makes grass grow, or a worm feel; much less are they able to give any new verdict on the nature and destiny of the human soul. —

Whatever the human soul may be, it is, our consciousness teaches us, the source of our intelligence and free will; it is that one principle which sees by means of the eyes, hears by the means of the ears, speaks by means of the tongue, etc.; it is that *one* life-giving force which unites all the manifold material component parts of the human body into *one* body, which it causes to grow, and which it governs and uses as an instrument, to accomplish the objects of its rational will. Modern scientists accept as a fundamental principle the indestructibility of matter and the persistence, or conservation, of force. They say, so far as observation reaches, we find that wherever matter or force undergo any changes, they are not destroyed, but only assume different modes of existence. The reason of this, no doubt, is, because what the Creator has once created, no creature is able to destroy, —although it may cause accidental changes. Therefore, modern scientists consider it an established truth that no particle of matter, and no minimum of force, in the existing order of nature, is destroyed by natural changes. Applying, then, this principle to the human soul modern scientists must say: Whatever the inmost nature of the human soul may be, it is no doubt a force, and the most noble to be found in the visible creation. Therefore, although it may cease to use the body as its instrument, yet this force itself must continue to exist, or, what is the same, this force must be immortal.

What will be the mode of existence or operation of the human soul after leaving the body, Modern Science is unable to decide. We will, therefore, have to look for instruction on this point to Divine Revelation and Christian Philosophy.

Before doing this, let us examine some points that may interest curious readers. What is the origin of the human soul? What is the final destiny of the life-giving principles of plants and animals?

Divine Revelation teaches, that God Himself created the soul of the first man. Gen. 2, 7. — It is the unanimous doctrine of prominent Catholic theologians, that this is the case with the soul of every man; it is directly created by God. Therefore we read Ecclesiastes 12, 7. of man: "The dust (body) returns into its earth, from whence it was, and the spirit returns to God, Who gave it."

Can modern scientists prove the contrary? Certainly not. They may assert that — *as far as observation reaches*—neither matter nor force are destroyed,—nor do they spring into existence from nothing; for all this can be done by the word of the Creator alone. Yet the origin of the human soul is one of those mysteries that lie far beyond the reach of scientific investigation. Prof. Asa Gray[5] observes that "increments of force by Divine action in time * * * could never in the least be known to science." Therefore, if Christian Philosophy teaches that God, in the order of nature established and every moment upheld by His will, is the Creator

_____
5. Natural Science and Religion, p. 96.

of every human soul, all what modern scientists can reply, is: Well, we cannot prove the contrary: for this is a matter that is beyond the reach of our means of observation.

Now, what becomes of the life-giving principles of animals and plants, when these cease to live? Although Divine Revelation has not deemed it necessary to give man a decisive answer to this question, and Modern Science cannot, it is nevertheless of interest to purely speculative philosophy. We shall therefore quote the following remarks—without deciding on them—from a work of the famous Spanish philosopher J. Balmes:[6] "What, then, will be the fate of these souls, or vital forces, when the organism to which they give life shall be destroyed? Shall they relapse into nonentity? * * * Shall they continue to exist? Shall they, perhaps, be destined to govern a new organism? * * * Before all, we must take notice, that we are here dealing in conjectures that have more reference to possibility than to reality. Philosophy may let us anticipate what *can* be, but not what *really is;* for the reality could be made known to us only by experience; but this is wanting in this case. * * * A created being is continually in need of being upheld in existence by the Creator. * * * When the object ceases for which something has been created, why could we not assume that this something will again be annihilated? I do not perceive that this would be contrary to Divine Wisdom or Goodness. Yet, let us assume that we do not wish

---

6. Fundaments of Philosophy, Vol. II., Chapt. 2.

to take refuge to annihilation, is it impossible that they (the life-giving principles of plants or animals) will continue to exist ?—For what purposes could they serve ?—I know not ; but the assumption may be admitted, that they might, again absorbed by the ocean of nature, be of some further use * * * Who has told us, that the vital principles of animals have no object any more to serve after the destruction of the organism which they enlivened ? Does the destruction of a plant annihilate all its vital forces ? * * * Who has told us that a vital force can be of use only as long as it exercises its influence on an object within the reach of our observation ?"

From this we see that Christian Philosophy may leave undecided the question, what becomes of the vital principles of plants or animals, after the destruction of the organisms to which they gave life.—Quite different, as explained before, it is with the soul of man, that free spiritual image of God. That this soul is immortal, is taught both by Divine Revelation and sound Reason ; and modern scientists have not the slightest solid argument to advance against this doctrine.

5. THE HUMAN SOUL AFTER THE DEATH OF THE BODY, AND THE DOCTRINE OF THE RESURRECTION.

Divine Revelation teaches that the soul of man is immortal ; and modern scientists have nothing to advance against this doctrine. Now, one may ask : What will be state of the human soul after the death of the body ?

In the first place, the soul will continue to exist only in its purely spiritual nature, similar to the angels. In the second place, the soul will continue to exercise its *spiritual*, but not its *sensitive* faculties, which man, in his present state of life, has in common with the more perfect animals. For, to exercise the sensitive faculties, corporeal organs, as eyes, ears, brain, etc., are necessary, and these perish with the body.

The great theologian and philosopher Suarez [1] teaches on this subject: "The soul separated (from the body) preserves the same subsistence which it had whilst in the body; nor does it change its entity (nature), but only its mode of existence; it thereby neither gains nor loses anything essential. * * * In the separated soul there remain the intellect, the will, and the power to move itself, as also the natural habits of the intellect and will, acquired in this life; moreover, the intellectual species (ideas,) but not the species (phantasmata) and habits that belong to the sensitive faculties. Therefore, the separated soul can understand and will; it knows all things which it knew in life, by means of the acquired ideas. It even understands better (melius intelligit) than it did when yet united to the body: for it understands independently of phantasmata."

St. Thomas [2] remarks that the separated soul is, to a certain extent, more at liberty in understanding (quodam modo est liberior ad intelligendum) in as far as, in this life, the purity of its intelligence is obscured by the grossness (gravedo) of the body, whose sensitive faculties present the phantasmata for the operations of the intellect

1. Theologiae Summa, Seu Compendium, Vol. I, Tom. III.
2. Summa Theologica, Pars I, quaestio 89, art. 2.

Although, then, the soul continues to exist without the body, and although it can continue to exercise its purely spiritual faculties in the future life, yet it must not be forgotten that the soul of man was not intended by the Creator to have a purely spiritual existence, like the angels,—but to be united with a body,—to be the connecting link between the material and purely spiritual worlds. Therefore, Christian theologians teach that the natural tendency or longing of the separated soul is, to be again united with the body (naturalis est animae conditio uniri corpori).

Divine Revelation teaches that this quite natural longing of the separated souls shall once be gratified;—they shall again be united with their bodies on the day of the general resurrection.

Now, what has Modern Science to advance against this doctrine? Certainly, no solid argument whatever. In the first place, Modern Science not only admits, but firmly holds that, in the existing order of things, no atom of matter nor the least amount of force can be destroyed by any created agency. Hence modern scientists will have to admit that, as far as scientific researches go, there is no reason to doubt that all atoms, or particles, of all the human bodies that ever existed, are still extant, and will continue to be so, as long as the present order of nature exists.

In the second place, modern scientists have no plausible reason for denying that God may, either directly or by the mediation of angels, cause all particles that ever belonged to a body, to be suddenly attracted by, or united to, the soul that once enlivened that body. For even in the present visible world we see that the Creator has endowed some created beings with the power of attracting others. It is well known that a powerful magnet will, as soon as it is brought sufficiently near, suddenly attract filings or pieces of iron, which may be hidden in dust or ashes. We know that in our solar system every celestial body, to some extent, attracts the other. If, then, material bodies, distant from one another many millions of miles, can exercise such a power of mutual attraction, why should not the Creator be able to endow also the human soul with the power of attracting suddenly, at the appointed moment of resurrection, all the particles that belonged to the body to which it had before given life ?

But some infidel scientist might suggest, it may happen that some particles become component parts of the bodies of different persons ; as, for instance, when cannibals eat the flesh of other people, or when men eat vegetables that have incorporated some material atoms once before belonging to the bodies of other persons.

We reply : In the first place, we know that no man incorporates in his bodily texture, all he eats.—Therefore, it might happen under God's providence that no man ever really and permanently incorporates in his body particles essentially belonging to the body of another. Secondly, physiologists teach that the component particles of our bodies are continually undergoing some changes—new ones are gained, old ones lost. Yet, our bodies remain substantially the same, whether some few particles are added or lost. Therefore, even if it should happen that some few particles would become successively incorporated with different bodies, Modern Science can yet advance no evidence to prove that not enough particles of every body would be left, to constitute again substantially the same former bodies. The famous theologian J. Perrone, S. J., observes :[3] " To constitute the identity of the body, the identity of every single molecule of matter is by no means required; * * * it suffices, that the essential parts be preserved, to restore the identical body."

From this we see, that Modern Science can advance no well founded argument against the doctrine of the Resurrection.

VII.   THE THEORY OF EVOLUTION IN THE ORGANIC WORLD.

HAVING examined what both Christian Philosophy and Modern Science teach concerning matter and life, concerning the principles of vegetative life, of animal sensation, and of human intelligence, or reason, and concerning the higher destiny and immortality of man : we will have to examine next these questions :—Did God create all the different species of plants and animals, that now exist, separately ? Or were the numberless different species now existing gradually evolved, under God's guiding providence, from some few, or perhaps only one directly created organism; as the leaves, flowers, and fruits of a tree are gradually evolved from a single given seed, or germ ?

3. Praelectiones Theologicae, Vol. V., Cap. VII.

Since, probably, no other scientific questions have, of late years, been discussed more throughout Christendom than these, they deverve our special attention.

It is beyond doubt true, that the immense majority of Christians have for many centuries been under the impression, that the various kinds of plants and animals, which have appeared on our Earth, were not gradually, in the course of ages, but quite suddenly called into existence by the Creator. Milton but expressed the then general belief of Christians, when he wrote:

> "The sixth, and of creation last, (day) arose
> With evening harps and matin, when God said:
> Let the earth bring forth soul living in her kind,
> Cattle and creeping things, and beasts of the Earth,
> Each in their kind! The Earth obeyed, and, straight
> Opening her womb, teemed at a birth
> Innumerous living creatures, perfect forms,
> Limbed and full-grown. Out of the ground up rose,
> As from his lair, the wild beast;
> The cattle in the fields and meadows green, etc."

Now, many modern scientists are decidedly in favor of the so-called "theory of evolution;" that is, they maintain, that the various species of animals and plants, now existing, have been evolved from a few, or a single, formerly existing, simpler forms; and these again have been evolved, as Huxley [1] thinks, from "undifferentiated protoplasmic matter which, so far as our present knowledge goes, is the common foundation of all vital activity."

Herbert Spencer is of the opinion that all phenomena of our visible universe are to be explained by the laws

---

1. Lectures on Evolution. Lecture I.

of evolution. He observes:[2] "From the earliest traceable cosmical changes down to the latest results of civilization, we shall find that the transformation of the homogeneous into the heterogeneous, is that in which progress essentially consists."

This theory of evolution has become exceedingly popular among modern scientists, especially through the influence of Laplace and Darwin. The former availed himself of this theory, to explain the origin of our present solar (or cosmic) system; the latter, to explain the origin of the now existing numberless species of plants and animals.

Having in a former article sufficiently explained the theory of Laplace, and its relation to Christian Doctrines, we shall now turn our attention to the theory of evolution as applied to the organic world.

In the conclusion of his famous work "On the Origin of Species" (Fifth Edition) Darwin remarks: "Authors of the highest eminence seem to be fully satisfied with the view that each species has been independently created. To my mind it accords better with what we know of the laws impressed on matter by the Creator, that the production and extinction of the past and present inhabitants of the world should have been due to secondary causes, like those determining the birth and death of the individual. When I view all beings not as special creations, but as the lineal descendants of some few beings which lived long before the first bed of the Silurian system was deposited, they seem to me to become

---

2. Progress: Its Law and Cause.

ennobled * * * It is interesting to contemplate a tangled bank, clothed with many plants of many kinds, with birds singing on the bushes, with various insects flitting about, and with worms crawling through the damp earth, and to reflect that these elaborately constructed forms, so different from each other, and dependent on each other in so complex a manner, have all been produced by laws acting around us. These laws, taken in the largest sense, being Growth with Reproduction; Inheritance which is almost implied by reproduction; Variability from the indirect and direct action of the conditions of life, and from use and disuse; a Ratio of Increase so high as to lead to a Struggle for Life, and as a consequence to Natural Selection, entailing Divergence of Character and the Extinction of less-improved forms. Thus, from the war of Nature, from famine and death, the most exalted object which we are capable of conceiving, namely, the production of the higher animals, directly follows.—There is grandeur in this view of life, with its several powers, having been originally breathed by the Creator into a few forms or into one; and that, while this planet has gone cycling on according to the fixed law of gravity, from so simple a beginning endless forms most beautiful and most wonderful have been, and are being, evolved."

These words explain what is called—Darwinism. We see that Darwin's theory rests on the following assumptions: 1. All existing species of organic beings have descended from a few, or perhaps only one form, into which the Creator originally breathed life. 2. The causes which effected the immense multiplication of organic *species* were: Variability in Growth and Re-production, a Struggle for Existence, in which the more favorably or perfectly constituted individuals survived, —or were Selected by Nature,—and the Inheritance of acquired qualities or perfections.

Here these questions arise: Does not Darwinism contradict the divinely revealed Mosaic Account of the Creation? Is Darwinism founded on undeniable facts?

The learned Jesuit-Father, F. von Hummelauer, observes:[3] "From what has been said one will easily see what the commentator of the biblical account of the Creation will think of Darwinism. We do not mean by Darwinism that gross error which sees in rational man no more than a highly developed ape. We here mean by Darwinism only that view, according to which all species of plants and animals have descended from a few, perhaps only one, original, perfectly simple form. It is, of course, not our object either to combat or defend this view (Darwinism); only that much we wish to establish that it, by no means, contradicts Genesis I. (that is, the divinely revealed Mosaic Account of the Creation). Darwinism lets only the first forms of life, or the one original form of all life, proceed immediately from the hand of God. The book of Genesis teaches that Adam saw (in a vision) all plants and animals "according to their kind" come into existence at the command of God. From this we have to infer only, that the book of Genesis ascribes the origin of all life to God, but whether this origin reaches back to the

3. Der biblische Schoepfungsbericht, p. 149.

Creator mediately or immediately, on this question we need not expect an explanation from Genesis, which (in describing the vision of Adam) had to pass over the intermediate links, if such existed."

From this we see that the Darwinian theory, provided it does not go so far as to degrade man to a species of apes, cannot be said to be inconsistent with the Mosaic Account of the Creation; although, of course, we do not intend to say that this Account teaches Darwinism. What we read in Genesis on this subject is: "God saith: Let the earth bring forth the green herb * * * and the fruit tree yielding fruit after its kind. * * * And it was so done." Gen. 1, 11. Again: "God said: Let the waters bring forth the creeping creature. * * * Let the earth bring forth the living creature in its kind. * * * And it was so done." Gen. 1, 20, 24. From this we see that the Mosaic Account teaches that the existing kinds of plants and animals *were brought forth by the Earth and the water at the command of God.* But whether these kinds of organic beings were brought forth at once, at the command of God, as most Christians believed,—or whether they came, under God's guiding providence, gradually into existence, as Darwin assumes,—on these points the Mosaic Account gives us no information. Consequently, on these points, there may be permitted, without danger of contradicting Divine Revelation, a great latitude of opinion among Christians.

We, therefore, need not be surprised to learn, that St. Augustin, the perhaps most profound and philosophic of the ancient doctors of the Church, in various works mentioned by Dr. Carl Guettler[4], expressed views decidedly favoring the theory of evolution. If one compares the views of St. Augustin with the speculations of Darwin, one might be tempted to look upon St. Augustin as the venerable teacher who advanced some grand comprehensive ideas, which his disciple Darwin has explained more in detail.

Dr. Carl Guettler remarks : "According to the view of this great doctor of the Church (St. Augustin), the entire inorganic and organic nature was created potentialiter atque causaliter (potentially and causably) at the same time with matter (Eccl. 18, 1), out of the informitas (formlessness) of which it (the entire inorganic and organic nature) was per temporis moras (in the course of time) developed. The world, at the beginning was like a seed which contained all the constituent parts of the future tree in itself invisibiliter (invisibly). Since St. Augustin considers also the body of man to have been created invisibiliter, potentialiter, causaliter (in(invisibly, potentially and causably) with the (original) matter,—and since he (St. Augustin) considers it (the human body) as a developed product of matter,—he, in principle, adhered to a theologically interpreted theory of evolution (that is, an evolution under God's will and guiding providence)."

These views of St. Augustin have by no means been repudiated as inconsistent with the Mosaic Account by the great Catholic theologians of the Middle Ages. St.

4. Naturforschung und Bibel, p. 145.

Thomas, commenting on Genesis 2, 4—5, observes:[5] "In the day, when God created Heaven and Earth, he created also every herb of the field, not actually, but before it grew upon Earth, that is, potentially."

From all this we see that the leading idea of Darwinism,—the gradual evolution of the existing species of plants and animals, need not shock any believer in the Mosaic Account, which ascribes only the *creation* of all things to God, but does not describe the exact manner of the creation.

Some short-sighted infidels have imagined, the Darwinian theory would help them to do away with the Creator. Darwin himself, by no means, considered the Creator superfluous; he was too intelligent a student of nature for that. He knew, without a Creator, it would be impossible to account for the origin of life. Therefore he assumed that life had "been originally breathed by the Creator into a few forms, or into one." Also the guiding providence of the Creator, in bringing forth, from quite simple forms of life, the countless now existing species,—has not been made superflous by Darwin's theory.—On the contrary, assuming the Darwinian theory to be correct, one will have to exclaim, as Prof. John Le Conte did in reference to the theory of Laplace: "How simple the means—how multiform the effects—how far-reaching and grand the design! How deeply they impress us with the wisdom, power, and glory of the Creator and Governor of the Universe!"[6]

5. Summa Theologica, Pars I, Quest. 74, Art. II.
6. The Popular Monthly Science, April, 1873, p. 655-6.

But Darwinianism must clear away some important difficulties, before it can be considered a well-founded theory, as we shall see from the following. Let us examine: First, the facts that seem to favor the Darwinian theory; secondly, the facts that seem to contradict the same; and thirdly, the facts which Darwin's theory cannot explain or prove.

Mr. George J. Romanes has published a little volume entitled "The Scientific Evidences of Organic Evolution;" and since he publicly declared (London, June 1, 1882) that "Darwin thought well of the epitome of his doctrine," we will briefly call attention to the groups of facts which Mr. Romanes advances in favor of Darwin's theory.

He observes: "I *first* take the argument from classification. Naturalists find that all species of plants and animals present among themselves structural affinities. * * * Our system of classification, therefore, may be likened to a tree, in which a short trunk may be taken to represent the lowest organisms which cannot properly be termed either plants or animals. This short trunk soon separates into two large trunks, one of which represents the vegetable and the other the animal kingdom. Each of these trunks then gives off long branches signifying classes, and these give off smaller, but more numerous branches, signifying families, which ramify again into orders, genera, and finally into the leaves, which may be taken to represent species. Now, in such a representative tree of life, the height of any branch from the ground may be taken to indicate the grade of organization which the leaves, or species, present; so that, if we picture to ourselves such a tree, we shall understand that while there is a general advance of organization from below upwards, there are numberless slight variations in this respect between leaves growing even on the same branch."

Mr. Romanes' second argument in favor of Darwin's theory is based on Morphology, or the structure of plants and animals. He remarks : " The theory of evolution by natural selection supposes that hereditary characters admit of being slowly modified, wherever their modification will render an organism better suited to a change in its conditions of life. * * * These changes would, in the first instance, begin to affect the least typical—that is, the least strongly inherited structures, such as the skin, claws, and teeth, etc. But as time went on, the adaptation would begin to extend to the more typical structures." In conformity with this view, Mr. Romanes tries to explain how terrestrial quadrupeds have become aquatic in their habits, etc. He also thinks this view to be confirmed by the so-called rudimentary structures ; of which he says : " Throughout the animal and vegetable kingdoms we constantly meet with organs which are dwarfed and useless representatives of organs which, in other and allied kinds of animals and plants, are of large size and functional utility. * * * How are they to be accounted for? Of course the theory of descent with adaptive modification has a delightfully simple answer to supply, viz., than when, from changed conditions of life, an organ which was previously useful becomes useless, natural selection combined with disuse and so-called economy of growth will cause it to dwindle till it becomes a rudiment."

The third argument in favor of Darwinism is taken from Geology. Mr. Romanes says: "Since the first dawn of life in the occurrence of the simplest organisms until the meridian splendor of life as now we see it, gradual advance from the general to the special—from the low to the high, from the few and simple to the many and complex—has been the law of organic nature."

As to the fourth argument Mr. Romanes observes: "The argument from Geology is the argument from the distribution of species in time. I will, therefore, next take the argument from the distribution of species in space—that is the present geographical distribution of plants and animals. It is easy to see that this must be a most important argument, if we reflect that as the theory of descent with adaptive modification implies slow and gradual change of one species into another, a still more slow and gradual change of one genus, family, or order, we should expect, on this theory, that the organic types living on any given geographical area would be found to resemble or to differ from organic types living elsewhere, according as the area is connected or disconnected with other geographical areas. And this we find to be the case, as abundant evidence proves."

A fifth argument, Mr. Romanes takes from Embryology. He says: "To economize space, I shall not explain the considerations which obviously lead to the anticipation that, if the theory of descent by inheritance is true, the life-history of the individual ought to constitute a sort of condensed epitome of the whole history of its descent. But taking this anticipation for granted, as it is fully realized by the the facts of Embryology, it follows that the science of Embryology affords perhaps the strongest of all the strong arguments in favor of evolution. * * * The higher animals almost invariably pass through the same embryological stages as the lower ones, up to the time when the higher animal begins to assume its higher characters. Thus, for instances, to take the case of the highest animal * * * (its) development begins from a speck of living matter similar to that from which the development of a plant begins. And, when (its) animality becomes established (it) exhibits the fundamental anatomical qualities which characterize such lowly animals as the jelly-fish. Next (it) is marked off as a vertebrate, but it cannot be said whether (it) is to be a fish, a snake, a bird, or a beast. Later on it is evident that (it) is to be a mammal; but not till still later can it be said to which order of mammals (it) belongs."

To these principal arguments, Mr. Romanes adds yet a few of minor importance. First, he thinks, this theory which was devised to explain facts of Biology, furnishes also an explanation of phenomena of Psychology. "This is especially the case with the phenomena of instinct, and in a lesser degree, with those of reason and conscience." Moreover, he considers it an argument in favor of the Darwinian theory, that not all structures and in-

stincts of animals are perfect in every way. Finally,
Mr. Romanes thinks, if "Divine Beneficence," and not
natural selection, had made the various species what they
are, not every species would be "for itself, and for itself
alone"—, but on the contrary, the various species would
have been so interrelated as to minister to each other's
necessities.

It is to be observed that these three minor arguments
in favor of Darwinism are not based on scientific
facts, but are merely deductions from Mr. Romanes'
speculative views, which are of no more value than those
of any other speculative person. Now, millions of intelligent persons may, and do differ with Mr. Romanes,
on these points. Yet, it is not our object to deal in
philosophical speculations, but to compare the results of
Modern Science with the Doctrines of Christianity.

Passing, then, over these points of minor importance,
let us briefly see what the opponents of Darwin's theory
reply.—To the principal arguments from Classification,
Morphology, and Embryology may be replied, that the
facts on which these arguments are based, may be explained, just as easily as on Darwin's theory, on the assumption that God created, according to a few "dominant
typical ideas," a vast series of organic, living species,
gradually filling up the immense chasm between the
lowest forms of visible existence—lifeless matter, and
man. This assumption would explain in a perfectly
satisfactory manner, why the numberless species are
united by a common bond, not of descent, but of "general
plans," or "dominant typical ideas."—Dr. Carl Guettler[7]

7. Naturforschung und Bibel, p. 165.

observes; "The general idea of likeness or similarity of appearance (Formen) is to be well disguished from genetic relationship (descent)."

Prof. Winchell[8] regards certain rudimentary structures as "premeditated intimations of the dominance of general plans." He calls attention to the fact, that there are many instances "in which the existence of organs in a rudimentary condition is historically *antecedent* to their existence in a fully developed condition." Thus, for instance, rudimentary lungs are found with tad-poles and various batrachians, as also with the gar-pike—a type of fish which existed an immense period of time before any air-breathing animal could live. Hence Prof. Winchell concludes : "On the hypothesis of an overshadowing plan of organic structure, framed by intelligence, carried into execution under the guidance of intelligence, behold how beautiful and how gratifying an explanation of all these rudimentary structures."

From this we see that the facts on which Mr. Romanes bases his arguments from Morphology, Classification, and Embryology, may, as least as easily as on the Darwinian hypothesis, be explained on the assumption that God created the various original species after a few "dominant typical ideas" according to a certain plan of succession.—Thus several of Mr. Romanes' arguments in favor of the Darwinian hypothesis lose their conclusiveness.

Let us next see what the opponents of Darwin's theory reply to Mr. Romanes' argument from Geology. It is

---

8. The Doctrine of Evolution, pp. 85-7.

true that, from the first dawn of life up to the last geological epoch, a gradual advance from less perfect to more perfect species of plants and animals can be traced. Yet, this only proves that more perfect species succeeded gradually, as the Earth became a fit abode for them, less perfect ones, but not, that the more perfect species have been gradually developed from the less perfect, as Darwin's theory would postulate.

One great stubborn truth stands in the way of the theory of organic evolution,—that notwithstanding variations, "we are ignorant of a single instance of the derivation of one good species from another," although our globe has been ransacked to find such an instance. No doubt, species may vary under certain conditions; but, as far as observation shows, "the divergent form, when relieved of physical constraint, rapidly reverts to its original type."

Dr. Carl Guettler[9] calls attention to the following facts : Numerous mummied specimens of oxen, cats, dogs, monkeys, crocodiles, and birds, from 3 to 4000 years old, have been discovered in Egypt. Yet, between these and the same still living species of animals not the slightest specific difference can be found. Some grains of wheat discovered with those ancient mummies were sown, and produced a variety of wheat exactly like one yet grown in Egypt. According to Agassiz, the formation of the coral-reefs of Florida lasted at least 70,000 years, and during that immense period of time the building polyps always remained specifically the same. The same is true of the conchylious mollusks of

9. Naturforschung und Bibel, p. 157—166.

the oecene, miocene, and pliocene tertiary period, which continue to exist yet to-day specifically the same as many thousands of years ago. The same is true of the various kinds of mollusks that have continued to exist unchanged since the remote Silurian Ages. The gar-pike, a type of fish which existed long before any air-breathed animal, is to-day yet just as its ancestors were countless thousands of years ago.—Numerous other instances of the persistence of specific types are mentioned by paleontologists.

If Darwin's theory were true, numerous fossil forms of transition from one species to another would have to be found in the geological strata ; but this is not the case. Wherever new fossil species appear, they do so quite abruptly.

To get over this difficulty, Darwin takes refuge to our limited knowledge of the various fossil species that may have existed without our knowing.

Dr. Guettler[10] replies : "In fact, this our ignorance (as to extinct fossil species) is by far not so great, as Darwin represents. We are acquainted with about 150,000 different species of fossil animals from the most different countries ; and of many species, thousands of fossil specimens are extant. If, then, really new specific forms had come into existence by gradual transition or change from pre-existing species, * * * some such forms of transition would necessarily have been preserved. * * * That just all the forms of

10. L. c., p. 159–60.

transition should have been destroyed by geological processes, is an assumption, the immense improbability of which is evident."

Now, if these facts from Geology and Paleontology prove anything, it is not in favor of the Darwinian theory, but directly destructive to it. These facts show that, so far as exact scientific researches reach back into past geological ages, organic species have, as long as they existed, always retained their specific identity.

There is yet one of the arguments which Mr. Romanes considers favorable to Darwin's theory, namely, the argument from geographical distribution. It is claimed, that the organic types living on any given geographical area are found to resemble or differ from organic types living elsewhere, in proportion as the area is connected or disconnected with the other geographical areas. The opponents of the Darwinian theory reply to this argument: All the facts on which it is based, may be satisfactorily explained on the assumption that various centres of creation existed,—where certain species appeared first. Now, from these centres the distribution of the various species through adjacent territories took place, in proportion to the greater or lesser facility with which the various species could migrate or be carried to other geographical areas. This explains why many organic types resemble another the more, as their geographical areas are more connected.

We have thus far seen, what arguments are advanced in favor of the Darwinian theory, and how the opponents of this theory reply to these arguments.

After impartially examining the arguments of both sides, the conclusion seems to be inevitable, that although the Darwinian theory may, and probably does, contain some valuable grains of truth,—yet, as commonly understood, it is far from being a well-founded theory; is it no more than a bold hypothesis.

Moreover, if some infidels talk or write, as if Darwin's theory had made the Creator superfluous, they only show their lack of reflection. Darwin himself knew better. Prof. Asa Gray who favors this theory, [11] declares: "Indeed, mediate creation is just what the thoughtful and thorough observer of the ways of God in Nature would expect, and what some of the illustrious of the philosophic saints and fathers of the Church have more or less believed in." This highly esteemed American scientist observes moreover: "Darwinism has real causes at its foundation, viz: the fact of variation, of the inevitable operation of natural selection, determining the survival only of the fittest forms for time and place. It is therefore a good hypothesis, so far. But is it a sufficient and a complete hypothesis?" Prof. Gray thinks not; for "natural selection" does not explain why lower forms should rise to higher ones; why simple ones should become complex; why protoplasm should change into a plant or animal; why a lower animal should become a more perfect one. "Natural Selection" cannot explain the origin of sensitiveness, consciousness, or intellect; it does not account for the origin or formation of any organ, as of the eye, brain, hand, etc.

---

11. Natural Science and Religion,

Hence infidels who imagine that Darwin's theory has made the Creator superfluous, are greatly mistaken.

From what has been said on Darwinism, we may draw the following conclusions: 1. Darwinism, within certain limits, does not contradict the divinely revealed Mosaic Account of the creation. 2. Darwinism is far from being an undoubtedly true theory. As Mr. Asa Gray observes : " From the nature of the case this conception can never be demonstrated ;" for all facts on which the arguments in favor of Darwinism, or the theory of the evolution of organic species, are based, can be explained also on the assumption that the various species were, according to comparatively few dominant plans, created successively.

Having reviewed the *pro* and *contra* of the famous hypothesis of the evolution of species in general, we will examine its application to man, the crown of the visible creation.

VIII. MODERN SCIENTIFIC VIEWS ON MAN.

1. MAN'S PLACE AND OBJECT IN NATURE.

MR. THOMAS H. HUXLEY observes :[1] " The question of questions for mankind—the problem which underlies all others, and is more deeply interesting than any other—is the ascertainment of the place which Man occupies in nature, and of his relations to the universe of things. Whence our race has come;

---

1. Evidence as to Man's Place in Nature.

what are the limits of our power over nature, and of nature's power over us ; to what goal we are tending; —are the problems which present themselves anew and with undiminished interests to every man born into the world."

God, who has created man, did not leave him in darkness as to his origin, his place in nature, and his final destiny. We read in the first book of the Bible : "God created man to his own image * * * : male and female, he created them. And God blessed them, saying : Increase and multiply, and fill the Earth and subdue it, and rule over the fishes of the sea, and the fowls of the air, and all living creatures that move upon the Earth. * * * Behold, I have given you every herb * * * and all trees, etc." Thus man was made by God lord over all upon Earth ; he was to be the crown of the visible creation ;—in him the visible, or material, and the spiritual worlds were united. As his body was taken from the Earth, so his soul was a breath, a spiritual image, of the Creator,—similar to purely spiritual beings called angels. The Psalmist 8, 4, 6, in the following words, briefly describes man's place in nature : "I will behold (O God) thy heavens, the works of thy fingers : the moon and the stars which thou hast founded. What is man, that thou art mindful of him ? Or the son of man, that thou visiteth him ? Thou hast made him a little less than the angels, thou hast crowned him with glory and honor : and hast set him over the works of thy hands."

Man, then, is placed between two worlds—the visible and the invisible. From the visible creation, he should rise with the wings of his spiritual faculties to the higher, invisible world,—to God Himself. Therefore Divine Revelation teaches: "The invisible things of Him (God), from the creation of the world, are clearly seen, being understood by the things that are made: His eternal power also and divinity." Romans 1, 20. And again the book of Wisdom teaches: "All men are vain, in whom there is not the knowledge of God; and who by these things that are seen, could not understand Him, that is, neither by attending to the works have acknowledged who was the workman; but have imagined either the fire * * * or the circle of stars or the sun and moon, to be the gods that rule the world * * * Let them know, how much the Lord of them is more beautiful than they: for the First Author of beauty made all those things. Or, if they admired their power and their effects, let them understand by them, that He that made them, is mightier than they: for by the greatness of the beauty, and of the creature, the Creator of them may be seen, so as to be known thereby."

Thus Divine Revelation teaches that all the visible universe is but a school in which man should learn to know his Creator; yet, not only—to know, but also—to love, and love Him above all things. The pious author of the Following of Christ, Book II, Chapt. 4, remarks: "If only thy heart were right, then every created being would be to thee a mirror of life and a book of holy teaching. There is no creature so little and so contemptible, that it sheweth not forth the goodness of God."

The visible creation, then, is to *lead man to God*, his Creator and the source of his eternal happiness. This is the reason why nothing created can satisfy the heart of man. Solomon, after attaining the summit of earthly glory and pleasures, felt compelled to confess : "I made me great works : I built me great houses, and planted vineyards : I made gardens and orchards, and set them with trees of all kinds * * * I heaped together for myself silver and gold, and the wealth of kings and provinces : I made me singing men, and singing women, and the delights of the sons of men, cups and vessels to serve to pour out wine ; and I surpassed in riches all that were before me in Jerusalem : my wisdom also remained with me. And I withheld not my heart from enjoying every pleasure * * * and when I turned myself to all the works which my hands had wrought * * * I saw in all things vanity, and vexation of mind, and that nothing was lasting under the sun." Eccl. 2, 4–11.

This, at the end, every one will have to confess, who has sought true and full happiness in anything created. The author of "Is Life Worth Living ?" truly observes: "The emptiness of things of this life, the incompleteness of even its highest pleasures, and their utter powerlessness to make us really happy, has been * * * a common place both with saints and sages."

Man is greater than the visible creation around him ; he has been created for God ; hence finite creatures cannot satisfy him.—The visible creation should be to him but a ladder on which he is not to rest, but to ascend to God, the Author of his eternal repose. Therefore St. Augustine exclaimed: "Fecisti nos ad Te, Domine! et inquietum est cor nostrum, donec requiescet in Te !"—

" Thou hast created us for Thee, O Lord ! and our heart is unquiet, until it will rest in Thee !"

This, then, is what Christian Doctrine teaches as to Man's Place and Object in Nature.

Now, what has Modern Science to say on this point ? Nothing,—absolutely nothing !—for such questions, on which God's Revelation alone can give full certainty, lie beyond the reach of scientific investigations ; neither the telescope, nor the microscope, nor chemical analysis, is here of any avail.

Let us turn our attention to another subject.

2. THE ORIGIN OF MAN.

It is well known what Divine Revelation teaches on this point. We read in the Book of Genesis, that, the Earth having been adorned with plants and animals of all kinds, God crowned the visible creation by making man according to His Divine image and likeness. "The Lord God formed man of the slime of the Earth : and breathed into his face the breath of life, and man became a living soul." Genesis 2, 7.

Some modern scientists, as Darwin, Haeckel, Huxley, etc., maintain that man was not created by God, but gradually developed from some lower animal species to which also the higher apes owe their descent. The infidel naturalist Haeckel has published what he imagined to be a complete genealogical tree showing Man's animal origin.

What are we to think of it ? Mr. Samuel Wainwright[1] hits the nail on the head with the following words :

1. Scientific Sophisms.

"The theory of man's ape-descent thus constructed is perfect—but it is in the air. It lacks but one thing to give it relevance; and that one thing is reality. Like the "chateaux en Espagne" of the penniless count, it exists only in the interested imagination of the pretender. Du Bois Reymond has incurred the bitter and unappeasable wrath of Haeckel by declaring this genealogical tree (Stammbaum) to be as authentic in the eyes of a naturalist, as are the (fabulous) pedigrees of the Homeric heroes in those of an historian."

Indeed, all talk about man's ape-descent has no other foundation than the gratuitous assumption of the truth of the Darwinian theory—carried to the utmost extremes.

On what doubtful foundations the whole Darwinian hypothesis is based, we have shown in a previous article; we shall now confine ourselves to showing briefly, how unfounded the application of this theory is to the origin of man.

Let us examine the arguments advanced by Mr. Thomas H. Huxley[2] in favor of the hypothesis of the development of man from the lower animals.

Mr. Huxley says : " It is a truth of very wide, if not of universal, application, that every living creature commences its existence under a form different from and simpler than that which it eventually attains. The oak is a more complex thing than the little rudimentary plant contained in the acorn ; the caterpillar is more complex than the egg ; the butterfly than the caterpil-

_____

2. Evidence as to Man's Place in Nature.

lar; and each of these things, in passing from its rudimentary to its perfect condition, runs through a series of changes, the sum of which is called its development. In the higher animals these changes are extremely complicated; but within the last half-century, the labors of such men as Von Baer, Rathke, Reickert Bishof, and Remak have almost completely unraveled them. * * * * It is a general law that the more closely any animals resemble one another in adult structure, the larger and the more intimately do their embryos resemble one another. * * * Thus, the study of development affords a clear test of the closeness of structural affinity, and one turns with impatience to inquire what results are yielded by the study of the development of man. Is he something apart? * * Or does he originate in a similar germ, pass through the same slow and gradually progressive modifications (as lower animals)? * * * Without question, the mode of origin and early stages of the development of man are identical with those of the animals immediately below him in the scale."

In addition to what has been stated in a preceding article as to the bearing of Embryology on the hypothesis of evolution, we quote the following words of the famous scientist Father Angelo Sechi, S. J.,[3] in reply to these remarks of Mr. Huxley. "It has been attempted to compute the number of oxygen-and hydrogen-atoms necessary for the formation of $\frac{1}{1000}$ of a cubic inch of water, and their number is estimated at about

---

3. "Die Groesse der Schoepfung," translated from the Italian.

3,900,000,000,000,000—three thousand and nine hundred billions. Yet water is one of the least composed bodies * * * From this we draw an evident conclusion: namely, we brand those impertinent and ignorant naturalists who, to uphold their assumption of the transmutation of species, claim that the original germs (Urzellen) out of which the living organisms develop themselves—are all the same, and who insist upon the fact that their (microscopic) instruments show no difference. The fools! they do not comprehend that the two original germs of which one produces, for instance, a bird, and the other, a fish, can and must be in the arrangement of their entrinsic parts just as different from each other, as are also the grown and fully developed animals. With even the strongest instruments, these original germs will always appear as small points, just as an elephant and a horse would appear to one who, from a plain, would see them on some distant mountain, as moving and hardly distinguishable points."

We therefore reply to Mr. Huxley's first argument, that he is greatly mistaken if he imagines that man originates "in a similar germ," like the lower animals. This germ is as different, and, since cause and effect must necessarily correspond, must be as different from the germ of any brute, as man is different from brutes. If Mr. Huxley does not *see* that difference, it by no means follows that the difference does not *exist*.

The next argument in favor of man's development from some lower animals, is taken from the similarities of bodily organization, or structure, found to exist between man and some animals. "Whatever part of the animal fabric," Mr. Huxley says, " — whatever series of muscles, whatever viscera might be selected for comparison, —the result would be the same,—the lower apes and the gorilla would differ more than the gorilla and the man."

That there exists a great similarity of bodily structure between man and some animals, was known for thousands of years before Huxley or any other modern naturalist called attention to that fact. It was well known to the earliest Christian teachers, that man, according to his body, belongs to the animal kingdom; —but, according to his spiritual, rational soul, he belongs to the higher, the spiritual world. Therefore man, as was shown in a previous article, is as distinct from the animal kingdom, as animals are from plants.

Moreover, as Dr. Albert Stoeckl[1] observes : " If the bodily structure of man shows any similarity to the bodily structure of an ape, does it therefore follow that man has descended from an ape ? By no means. Only then such a conclusion could be drawn, if by other empiric facts it could be proved that man could have received such a body *only* by having descended from an ape. But such facts do not exist."

The bodily similarity between man and the most perfect animals, is easily accounted for on the Christian doctrine of man's place in nature.—Man was to be the centre in which the lower, material, and the higher, spiritual world were to meet ; man was to be the connecting links between these two worlds. Therefore, as according to his rational soul he was created similar to

---

4. Der Materialismus, p. 65.

the angels immediately above him in the scale of created perfection, so also, according to his body, he was created by God similar to the animals, immediately below him.

And now, let Mr. Huxley and other naturalists exert all their wits, to show the similarity of bodily structure between man and the more perfect animals; they can never refute the Christian doctrine of man's place in nature.

St. George Mivart,[5] after describing minutely the anatomical structures of apes and man, remarks : "To return from this subordinate question, it may be asked, 'What is the bearing of all the foregoing facts on the origin and affinities of man?'" He answers: "In nature there is nothing great but man. In man there is nothing great but mind. We must entirely dismiss, then, the conception that anatomy by itself can have any decisive bearing on the question as to man's nature and being as a whole."

But what does Geology, or rather Paleontology, teach on the subject of man's descent ? Mr. Huxley, towards. the end of his publication on Man's Place in Nature, after examining the cases of the Engis and Neanderthal skulls, frankly admits : "In conclusion, I may say that the fossil remains of man hitherto discovered do not seem to me to take us appreciably nearer to that lower pithecoid (ape-like) form, by the modification of which he (man) has, probably, become what he is." With these words, Mr. Huxley admits that there exists

5. Man and Apes, p. 187-8.

no proof whatever from Geology, that there ever existed, what is called the missing link between man and ape-like animals.

This fact is candidly admitted by the leading champions of man's animal descent. Mr. Haeckel's[6] prolific imagination has even invented a plausible reason to explain why the "missing link" has not yet been found, and why it will, probably, not be found for a long time to come. He asserts quite coolly: "There exist a number of indications which point to the fact that the original home (Urheimath) of man was a continent now sunk beneath the surface of the Indian Ocean."

What a pity for Haeckel & Co., that they will have to wait for the discovery of the "missing link," until it will please that continent to peep again out of the Indian Ocean !

Yet, whatever may happen in the future, Mr. Haeckel is candid enough to admit[7] for the present: "Of the hypothetical original man (Homo primigenius !), who * * * developed himself during the Tertiary period from anthropoid apes, we are as yet acquainted with no fossil remains."—Mr. Haeckel, then, admits that there exists no tangible proof that such an original (!) man ever existed.

Mr. Darwin[8] also admits that there exists—"the great break in the organic chain between man and his nearest allies, which cannot be bridged over by any extinct or living species (known)." He adds : "Nor should it be

6. Natuerliche Schoepfungsgeschichte, 4th Edition, p. 619.
7. L. C., p. 620.—
8. The Descent of Man. Part. I, Chapt. VI.

forgotten that those regions which are the most likely to afford remains connecting man with some extinct ape-like creature, have not as yet been searched by geologists." Hereby Mr. Darwin also candidly admits that up to the present time the missing link between man and some ape-like creatures has not yet been found—outside of some prolific imaginations.—But, perhaps, some future, yet unborn geologist *may* find it !—Such as have patience, are kindly requested to wait for this discovery ! !

From all what has been said on the subject, the reader will see that the so-called descent of man from some ape-like creature, is only a mere assumption—without any solid facts in its favor,—and with serious objections against it.

Among others, Mr. Alfred Russell Wallace,[9] a prominent naturalist and champion of the Darwinian theory, has shown that there exist important "limits of Natural Selection as applied to man." Mr. Wallace is of the opinion that Darwinism can explain no more "the development of man from lower animals," than it can explain "the origin of sensation or consciousness."

It would be an easy matter to quote many similar statements made by prominent naturalists ; yet, to avoid tiring the reader, we will add only the following words of the famous French naturalist A. De Quaterfages:[10] "As for us, gentlemen, we do not pretend to be either theologians or philosophers. We are exclusively men

---

9. The Action on Natural Selection on Man.
10. The Natural History of Man, New York, 1875, p. 87.

of science ; we have, then, to disturb us, only the truths
of science.   It is in the name of these truths that I have
had to recognize the weakness of science, to say: Whence
comes man ? But, in the name of scientific truth, I can
affirm that we have had for ancestor neither a gorilla,
nor an orang-outang, nor a chimpanzee ; no more than a
seal or a fish, or any other animal whatever."

Modern Science, then, has no well-founded fact to advance against the doctrine of Divine Revelation, that
"God created man to his own image ;" that "the Lord
God formed man of the slime of the Earth : and breathed
into his face the breath of life." Gen. I. & II.

Here we may be permitted to call attention to some
purely speculative questions suggested by the scientists
Mivart and Asa Gray—in reference to the origin of
man's body—and man's immortal soul.

Prof. Mivart, a Catholic scientist of England, as Dr.
Schaefer[11] states, has expressed the opinion that it
would not be inconsistent with any clearly established
doctrine of Divine Revelation, to maintain that the body
of the first man was gradually perfected by evolution,
or development, from some lower animal species,—and
that after this body had reached the perfection contem-
plated by the Creator, it was endowed by Him with the
spiritual and immortal soul. Also in this case, the body
of man could be said to have been taken from "the
Earth," though not directly but mediately.

What are we to think of this opinion ? In the first
place, it is a purely speculative matter, not a question

----

11.   Bibel und Wissenschaft, Muenster, 1881, pp. 277-8.

concerning results of scientific researches. Secondly, Prof. Mivart's opinion has not been censured by the proper ecclesiastical authorities, although some theologians consider it inconsistent with the Mosaic account of the creation of man. Thirdly, if Dr. Carl Guettler [12] is not mistaken, it would seem that even the great teacher St. Augustine did consider this opinion as in harmony with Christian Doctrines. Dr. Guettler, after referring to several works of the great teacher, observes : "Indem Augustinus auch den koerperlichen Menschen *invisibiliter, potentialiter, causaliter* mit der Materia sich erschaffen denkt und ihn als ein Entwicklungsproduct hinstellt, pflichtet er in Princip einer teleologisch interpretirten Pithekoidentheorie bei."

After carefully considering both sides of the question, I for one would not venture to declare Prof. Mivart's opinion inconsistent with any Christian Doctrine, although great theological authorities [13] decidedly reject the same.

Another novel opinion, as to the origin of man's immortal soul, has been advanced by Prof. Asa Gray. [14] In order that this eminent scientist may not be misunderstood on this point, I call special attention to the following statements : Prof. Gray expressly declares : 1. "When the naturalist is asked, what and whence the origin of man, he can only answer in the words of Quaterfages and Virchow, 'We do not know at all.' We have traces

---

12. Naturforschung und Bibel, Freiburg, 1877, p. 145.
13. See H. Hurter S. J.: Theologiae Dogmaticae Compendium, vol. II 1878, p. 180.
14. Natural Science and Religion, New York, 1880, pp. 100–6.

of his existence up to and even anterior to the latest
marked climatic change in our temperate zone : but he
was then perfected man ; and no vestige of an earlier
form was known. The believer in direct or special creation is entitled to the advantage which this negative
evidence gives." 2. "Sober evolutionists do not suppose that man descended from monkeys. The stream (of
descent) must have branched too early for that." 3. Prof.
Gray admits the great superiority of man over animals.
He says : "A being who has the faculty—however bestowed—of reflective, abstract thought superadded to all
lower psychical faculties, is thereby *per saltum* immeas-
urably exalted * * * None of us have any scientific or
philosophical explanation to offer as to *how* it came to
be added to what we share with the brutes that perish ;
but it puts man into another world than theirs, both
here, and—with the aid of some evolutionary ideas, we
may add—hereafter." *Now comes the main point*, namely
how Prof. Gray, according to the theory of evolution,
wishes to explain *the origin of man's immortal soul.* He
says : " Now see how evolution may help you; in its conception that, while all the lower serves its purpose for the
time being, and is a stage toward better and higher, the
lower sooner or later perish, the higher, the consummate,
survive. The soul in its bodily tenement is the final
outcome of Nature. May it not well be that the perfected soul alone survives the final struggle of life, and indeed 'then chiefly lives,' because in it all worths and
ends in here ; because it only is worth immortality, because it alone carries in itself the promise and potentia-

lity of eternal life ! Certainly in it only is the potentia-
lity of religion, or that which aspires to immortality."

This hypothesis would seem both ingenious and consistent with Christian Doctrines,—if we could assume,
as some early Christian teachers did, among them Tertullian,[15] that " as the body is born of the body, so the soul is
born of the soul." But for theological and philosophical
reasons, which it is not our object at present to explain,
this opinion of Tertullian and others has been rejected
by all prominent modern Catholic theologians, who
teach that the human souls, being spiritual, immortal
substances, cannot be propagated by generation, or be
the result of a process of organic evolution, but must
be called into existence by direct creative acts of God.

Modern Science has no proved fact, to show the
contrary. Prof. Gray[16] observes : " Increments of force
by Divine action in time, * * * if such there be,
could never in the least be known to science." In a
similar manner we say, The creation of human
souls by direct Divine action in time, as taught by
Catholic theologians, is beyond the reach of human
observation, and can hence never in the least be known
to science.

3. ANTIQUITY OF MANKIND.

Mr. Huxley[1] observes : " If any form of the doctrine
of progressive development is correct, we must extend
by long epochs the most liberal estimate that has yet
been made of the antiquity of man."

15. H. Hurter, S. J., in the work mentioned before, p. 343.
16. Natural Science and Religion, p. 96.
1. Evidence as to Man's Place in Nature.

Some infidel scientists entirely overlook Mr. Huxley's "If"—and talk of an enormous antiquity of mankind, as of a matter of fact. Mr. Nathan Allen [2] says of some scientists: "If these authorities are right, then at a period earlier than 200,000 years since, Europe was peopled by palaeolithic men."

Yes—"if these autorities are right," then, etc. But just *this* will appear very improbable to any one who carefully examines the foundations on which the assertions of an enormous antiquity of mankind are based.

It will not do to say, it *must* have taken countless thousands of years, before man could have become so perfectly developed, as he is. For, as was shown before, there is no proof extant, that man ever was "developed," as Haeckel & Co. dream.

Let us see on what foundations an estimate as to the antiquity of mankind can be based:

1. What does reliable history teach in reference to the antiquity of man? How many thousands of years before Christ do reliable histories of nations reach back into antiquity?

Every student of history knows that the reliable history of the ancient Romans reaches no further back into antiquity than at most 750, of the ancient Greeks, hardly 1000 years before Christ.

As to the antiquity of other nations, the learned English historian Prof. George Rawlinson, of Oxford, observes: [3] "Cuneiform (inscriptions) scholars confidently

2. The Popular Science Monthly, November, 1882, p. 97.
3. The Origin of Nations. Chapt. IX.

place the beginnings of Babylon about B. C. 2300, of Assyria, about B. C. 1500. The best Aryan scholars place the dawn of Iranic civilization about B. C. 1500, of Indic about B. C. 1200. Chinese investigators can find nothing solid or substantial in the past of the "Celestials" earlier than B. C. 781, or at the furthest B. C. 1154. For Phoenicia the date assigned by the latest English investigator is "the sixteenth or seventeenth century before Christ." The researches of Dr. Schlieman in the Troad give indications of the existence of a low type of civilization in that region, which may reach back to about B. C. 2000. In the rest of Asia Minor we have no certain knowledge of any civilization that has a greater antiquity than about B. C. 900. In Europe, the simple and incipient civilization delineated by Homer must have commenced as early as the Trojan epoch, which is probably about B. C. 1300—1200. No other European civilization can compete with this, the Etruscan not reaching back further than about B. C. 650 or 700, and the Celtic, such as it was, being really subsequent to the occupation of England by the Romans. *A consensus of savants and scholars almost unparalleled* limits the past history of civilized man to a date removed from our own time by less than 4400 years, *excepting in a single instance.*" And the single instance referred to is Egypt, with respect to whose antiquity savants are at variance, because the chronological data concerning the ancient history of Egypt are a mass of confused statements, so that one often cannot distinguish between what has happened contemporaneously, and what has happened successively. Mr. Mariette, quoted by Mr. Rawlinson, remarks : "The greatest obstacle to establish a regular Egyptian chronology is the circumstance that the Egyptians themselves *never had any chronology at all*."

We, therefore, need not be surprised that some placed the reign of Menes, who is considered to have been the first king of Egypt, at as remote a date as about B. C. 5000.—But others are in favor of a by far later date. Mr. Rawlinson states [4] that the following authorities gave these dates as 'the time of Menes' reign : Dr. Brugsch, Director of the Museum of Antiquities in Berlin, 4400 B. C. ; Dr. Lepsius, 3892 B. C. ; Baron Bunsen, 3059 B. C. ; Reginald Stuart Poole, 2717 B. C.; Sir Gardner Wilkinson, "who," as Mr. Rawlinson observes, "on the whole, must be regarded as the greatest of English Egyptologers," considers as the proximate date of the accession of Menes—the year B. C. 2691.

Mr. Rawlinson himself considers about B. C. 2450 as the time of the establishment of a settled monarchy, and with it, of civilization in Egypt. "This view," he observes, "appears to us to be more in accordance than any other with the general facts of oriental history and chronology. Its compatibility with the chronology of the Bible will be evident, if it be born in mind that, according to the Septuagint version, the date of the Deluge was certainly anterior to B. C. 3000."

From all this we see that the reliable results of historical investigations concerning the most ancient history of mankind perfectly agree with what the inspired Written Word of God, the Bible, relates up to the time of the Deluge.

---

4. The Origin with Nations. Chapt. II.

Let us next see what Modern Science can tell us of the history mankind before the time of the Deluge (about B. C. 3000).

Mr. Ingersoll says[5]: "We know, if we know anything, that men lived in Europe with the hairy mammoth, the cave-bear, the rhinoceros, and the hyena. Among the bones of these animals have been found the stone hatchets and flint arrows of our ancestors. In the caves where they lived have been discovered the remains of these animals that have been conquered, killed and devoured as food, hundreds of thousands of years ago. If these facts are true, Moses was mistaken."

Exactly so, Mr. Ingersoll, "If." But how does Mr. Ingersoll know, that men, no less than "hundreds of thousands of years ago," have killed animals? To him and a few scientists this fact seems to be evident, *because* men lived with the hairy mammoth, the cave-bear, etc. Among the bones of these animals, stone hatchets and flint arrows have been found; and, *therefore*, man *must* have lived "hundreds of thousands of years ago!"—And such palpable nonsense—that great light of American unbelief is not ashamed to publish.

Since such silly assertions as to the antiquity of mankind are often made, it may not be out of place to examine briefly their supposed foundations. If any reader wishes to study more on this subject, we advise him to read "The Recent Origin of Man, as Illustrated by Geology and the Modern Science of Prehistoric Archaeology. By James C. Southall, 606 Pages. Published: Philadelphia. By J. B. Lippincott & Co., 1875." In this

5. Some Mistakes of Moses, p. 100.

work the reader will find about all facts worth mentioning bearing on this question.

That man was coeval in Europe with some now there extinct animals (as, for instance, two species of elephants, two species of rhinoceros, one species of hippopotamos, the cave-bear, cave-lion, and cave-hyena, and some other animals) is, as Dr. Guettler[6] declares, now generally admitted.

Yet, that is no proof for an enormous antiquity of mankind, as Mr. Ingersoll and others seem to imagine. As Dr. Guettler remarks,[7] these animals may have existed for thousands of years *before* man's creation ; but the last ages of their existence may have been coeval with the appearance of man, to whom the destruction of many of these animal species may be ascribed. No sound reasons can be assigned, why these now extinct species of animals may not have yet existed and been hunted by men in the interior of Europe at the time when the Phoenicians traversed the Mediterranean Sea in every direction, when the Greeks besieged Troy, or Moses led the Israelites through the desert. The gradual disappearance of certain species of animals in some countries, within the strictly historic times, is nothing new.

It is well known how rapidly some animals are disappearing in the United States ; as the buffalo, the bear, the wolf, the elk (which still existed 1832 in Wisconsin) the antelope, the wild turkey, the otter, the beaver, the moose, the bison, etc. Even at the time of Aristotle and Herodot, the lion existed yet in Greece and Macedonia.

---

6. Naturforschuug und Bibel, p. 260.
7. L. C., p. 265.

The moa and the dodo disappeared in modern times. The elephant and rhinoceros are fast disappearing in India. The brown bear still existed in Belgium in the Middle Ages. The urus existed in Germany even as late as the Middle Ages; nevertheless, its remains are also found in the famous "bone-caverns." The same may be said of the aurochs, etc.

If, therefore, now and then, some traces of human remains are discovered mixed up with the bones of now extinct species of animals, there is yet no conclusive reason for assuming that man *must* have lived many countless thousands of years ago. There is no evidence whatever to show that any of those animals *must* have lived with man more than two thousand years before Christ.

Another unfounded assumption it is, because in *some* countries utensils or instruments of stone, bronze, and iron, have been found, that the history of mankind in general is to be divided into three corresponding epochs —each of them of an enormous duration. Whilst the Greeks and Romans were using iron weapons, numerous inhabitants of northern and central Europe still remained satisfied with weapons of stone, horn, or wood. Some uncivilized tribes, as many of our Indians, continued, even up to our own century, to live in the so-called "stone-age,"—of which some dream that it *must* have been many thousands of years ago.

Also the so-called lake-dwellings and caves, in which men formerly lived in Europe, do not prove anything in favor of a higher antiquity of mankind than reliable history admits. Probably, the famous lake-dwellings

are no older than a few centuries before Christ ;—and
" the ordinary cave-man," as even Mr. Allen Grant
admits,[8] " was superior in ingenuity and mental power
to nine out of ten among our own modern savages, and
quite equal to the run of our own laboring classes."

What solid reason, then, is there for assuming that
their antiquity reaches beyond even one thousand years
before Christ ?

Some other so-called arguments in favor of a very
remote antiquity of mankind, taken from the *duration*
of the formation of beds of peat, of geological
changes, etc., are not worth discussing, for they are
based on the totally false assumption that these formations, or changes, must have taken place " gradually and
uniformly ; " whereas they often occurred suddenly and
irregularly.[9]

There are yet these questions to be examined : Did
man live in the so-called Ice-Age ? If so : How many
thousands of years ago has this been ? We may call these
questions "most interesting," as far as the antiquity of
mankind is concerned ; for many so-called arguments in
favor of an enormously high antiquity of mankind,
depend on the solution of these questions.

To treat these questions clearly and briefly, we shall
examine : 1. What do we mean by the " Ice-Age " or
" Glacial-Period?" 2. How long ago was it ? 3. Did
man exist in or before it ?

What is meant by the Ice-Age ? Geologists assert
that after such animals as the horse, the camel, the

---

8. The Popular Science Monthly. November, 1882, p. 95.
9. Evidences of Religion, by Louis Jouin, S. J., p. 146.

elephant, the bear, etc., had been in existence for a long while, the high northern latitudes which were covered with the warm waters of the ocean experienced a remarkable uplifting which caused great climatic changes. Glaciers covered lands which before had enjoyed a pleasant climate,—and, in America, carried pieces of rocks from northern regions, and copper from Lake Superior, etc., as far down south as to the Ohio River.

Some think the Ice-Age came on gradually ; whereas others are of the opinion that it made its appearance quite suddenly. Some again assume two distinct Ice-Ages ; whereas others are inclined to think that the respective phenomena can be explained on the assumption of but one such period.

How long ago was the Ice-Age ? How little so-called Modern Science knows on the subject may be inferred from the following statements of modern scientists : [10] Mr. Croll estimates the beginning of the Glacial Period at two hundred and forty thousand years ago, and the period itself, as having lasted one thousand and six hundred centuries ! Other geologists estimate the Ice-Period to have been one billion two hundred and eighty million years ago ! ! Sir C. Lyell, at first, thought the Ice-Period to have been about eight hundred thousand years ago,—but, after more mature reflection, he came to the conclusion that two hundred thousand years would suffice.

Other authorities on Geology think, from what is known with certainty, that the Ice-Age, or Glacial Period, may have been *but a few thousand years ago.*

---

10. See Recent Origin of Man, by James C. Southall, pp. 47, 263, 495.

In the second volume of the Transactions of the Chicago Academy of Science, a paper by Prof. Andrews was published on "The North-American Lakes considered as Chronometers of Post-Glacial Time," which seems to be a complete refutation of the prevailing opinion as to an extremely remote period of the Ice-Age.

Dr. Andrews comes to the conclusion that "the total time of all the deposits, since the Ice-Age, appears to be somewhere between five thousand three hundred, and seven thousand five hundred years."

This would bring the time of the Ice-Age, or Glacial Period, and the Biblical Deluge closely together. Dr. Lorinser [11] does not consider it impossible, or improbable, for all we know, that the Biblical Deluge and the Geological Diluvium caused by the Ice-Age, had some intimate connection, or were, perhaps, identical.

The last question to be answered, is : Did man exists at the time of the Ice-Period ?

Dr. Guettler [12] says : "Whilst Perty, Schleiden, Unger, and Buechner, assert that men lived in Switzerland before the Ice-Period, Vogt, Lyell, and Pfaff, unanimously declare : "We have everywhere found evidences of the appearance of man *only after* the formation of the glacial till—in Scandinavia, England, and Switzerland."

Thus we see that geologists disagree amongst themselves, as to whether, or not, any traces of man have been found, that are older than the Ice-Period.

On reviewing what has been said in this article in reference to the Antiquity of Mankind, we come to the

11. Geologie und Palaeontologie, p. 242.
12. Naturforschung und Bibel, p. 264.

following conclusions : 1. All undoubtedly established historical facts are perfectly consistent with what the Bible relates since the time of the Deluge. Mr. G. Rawlinson [13] remarks : "According to the Septuagint version (of the Bible), the date of the Deluge was certainly anterior to B. C. 3000." And according to the same authority, there is no proof extant that the most ancient monarchies—of Egypt and Babylon—were founded any earlier than about 2,500 years before Christ. 2. As to the existence of men before the time of the Deluge, or before the origin of those two most ancient monarchies, Modern Science knows nothing certain.

We may therefore safely assert that Modern Science does not contradict the Bible as to the Antiquity of Mankind.

4.   THE DELUGE.

The traditions of the most ancient nations, with remarkable unanimity, relate that a great Deluge once swept nearly all mankind out of existence. Such traditions were met with not only among the ancient Greeks, Persians, Chinese, Babylonians, etc., but also among the Mexican and other American Indians.[1]

What these various traditions describe in a confused and distorted manner, is clearly stated by the Bible. We read : " The water was fifteen cubits higher than the mountains which it covered. And all the flesh was destroyed that moved upon the Earth, both of fowl, and

---

13. The Origin of Nations. Chapt. II.
1. See The Recent Origin of Man, pp. 34-36; and, Die Traditionen des Menschengeschlechts, von Dr. Heinrich Luecken, pp. 180-267.

of cattle, and of beasts, and of all creeping things that creep upon the Earth: and all men." (Genesis 7, 20–21.)

What were the causes of this Deluge? In the first place, the will of God Who intended to punish the crimes of mankind. As to the secondary causes which God used as means, the well known French scientist Louis Figuier remarks: [2] "All the particulars of the Biblical narrative * * * are only to be explained by the volcanic and muddy eruption which preceded the formation of Mount Ararat. The waters which produced the inundation of the countries proceeded from a volcanic eruption accompanied by enormous volumes of vapour, which in due course became condensed and descended on the Earth, inundating the extensive plains which now stretch away from the foot of Ararat. The expresssion 'the Earth,' or 'all the Earth' as it is translated in the Vulgate, which might be implied to mean the entire globe, is explained by Marcel de Serres, * * * and other philologists, as being an inaccurate translation. He has proved that the Hebrew word 'haarets,' incorrectly translated 'all the Earth,' is often used in the sense of *region* or *country*, and that, in this instance, Moses used it to express only the part of the globe which was then peopled, and not the entire surface. In the same manner 'the mountains' (rendered 'all the mountains' in the Vulgate), only implies all the mountains known to Moses. Similarly, M. Glaire, in the 'Chrestomathie Hebraique,' * * *

2. The World before the Deluge. pp. 481-2.

quotes the passage in this sense : 'The waters were so prodigiously increased, that the highest mountains of the vast horizon were covered by them ;' thus restricting the mountains covered by the inundation to those bounded by the horizon. Nothing occurs, therefore, in the description given by Moses, to hinder us from seeing in the Asiatic Deluge a means made use of by God to chastise and punish the human race, then in the infancy of its existence, and which had strayed from the path which He had marked out for it. It seems to establish the countries lying at the foot of the Caucasus as the cradle of the human race ; and it seems to establish also the upheaval of a chain of mountains, preceded by an eruption of volcanic mud, which drowned vast territories entirely composed, in these regions, of plains of great extent." Thus far Mr. Figuier.

This view may, at least substantially, be adopted by believers in the Bible.[3] According to this view, the Deluge would have covered *only* those regions of western Asia, which were then inhabited by *mankind*; whilst such distant countries, as Australia, America, etc., would not have been reached by it. In these countries, animals and plants would not have been interfered with.

Some believers in the Bible even go so far as to assume that, besides the family of Noah, some Mongolian or Ethiopian families which had migrated into distant regions, perhaps beyond the Himalaya Mountains, or to Central Africa, may have escaped the Deluge. The Jesuit Father Bellynk (Etudes religieuses, 1868, I, p. 578) ob-

---

3. See Naturforschung und Bibel, von Dr. Guettler, pp. 266-78.

serves[4] that "he would not censure such as think that
this hypothesis will one day become accepted." Yet,
some authorities are not inclined to favor this opinion.

By keeping these leading ideas in mind, it will be
easy to answer the silly objections advanced by Mr. Ingersoll and other unbelievers against the biblical narrative of the Deluge.

5.   THE LONG LIFE OF THE ANCIENT PATRIARCHS.

Mr. Ingersoll declares in his "Interviews": "It is unscientific to say that people at one time lived to be
nearly a thousand years of age." To Mr. Ingersoll a
great many things may appear to be "unscientific",
because his limited intelligence cannot comprehend
them.

To him it might appear "unscientific" to maintain
that such enormous animals as the Ichthyosaurus and
Plesiosaurus lived in remote geological ages; and yet
some of their skeletons have been found. Mr. Ingersoll
might call it "unscientific" to say that formerly tropical
plants, like the palms, flourished within the Arctic Circle,
where now only moss and low shrubs thrive; and yet remains of them have been discovered there. Mr. Ingersoll
might call it "unscientific" to claim that great numbers
of mastadons once lived in the now continually ice-cov-
ered northern Siberia; and yet their tusks are found
there by the hundreds.

Thus Mr. Ingersoll may also boldly assert that it is
"unscientific" to say that in remote ages men lived nearly

---

4. See Naturforschung und Bibel, von Dr. Carl Guettler, pp. 266-78.

a thousand years ; and *yet* the most ancient traditions of mankind relate it to have been a fact. Not only Moses relates it, but also the Egyptian Manetho, the Chaldean Berosus, the Phoenian Mochus, the Greeks Hekataeus, Akusilaus, and others. The ancient Persians, Chinese, and Hindus, insinuate the same fact.[1]

Who is a greater authority on this point, Mr. Ingersoll—living several thousands of years after the events referred to—or the most ancient nations that were near the times of those long-lived Patriarchs ?

If Mr. Ingersoll cannot comprehend *how* these could live so long, he may console himself with the words of Shakespeare :

"There are more things in Heaven and on Earth,
Than are dreamt of in your philosophy."

It is well known that if tropical plants are transported to colder or less favorable climates, they will grow weakly, and gradually deteriorate. In a similar manner it seems to have been with mankind. As long as mankind lived in a climate most favorable to its constitution, probably near the former Paradise, men attained an enormous age ; but after the climate had become considerably changed, perhaps by the Deluge, the average age of man commenced to decrease.

6. THE SPECIFIC UNITY OF MANKIND.

It is one of the fundamental doctrines of Christianity, that all men have descended from Adam.

There was a time when the doctrine of the origin of all mankind from some common ancestors was ridiculed by

1. See: Die Traditionen des Menschengeschlechts von Dr. Luecken.

unbelievers. Voltaire thought it could be believed only by blind persons, or by such as never had seen people of different races. Darwin and others have satisfactorily shown that great physiological differences may arise within the same organic species. Now-a-days, when some naturalists in all earnest maintain that even certain apes and man have descended from some common stock, it is seldom to find any scientist of note, who denies the possibility of the origin of all existing races of men from some common parents.

Mr. John Brocklesby[1] remarks on the subject : "The anatomical structure and physical constitution of man point decidedly to the unity of the race : the true skin * * * is alike in structure in all nations ; there is the same general coincidence in respect to the age when manhood is attained, and to the period when life begins to decline ; all races are subject to similar diseases, modified by varied climatic influences ; and the length of life is, on an average, the same under similar conditions of existence. The resemblances which exist throughout the languages and dialects of the world attest the same fact ; for the Indo-European group of nations * * * are all bound to each other by the affinity of their languages,—though they possess every shade of color belonging to the human race, ranging from the fair and ruddy complexion, through the swarthy and olive, to the deep black."

The learned Pritchard, author of the "The Physical History of Mankind," observes on the same subject : " We contemplate among all the diversified tribes

<hr />

1. Elements of Physical Geography, p. 147.

of the human race the same internal feelings, appetencies, and aversions ; the same inward convictions, the same sentiments of subjection to invisible powers, and of accountableness, more or less developed, to unseen avengers of wrong, from whose tribunal men cannot, even by death, escape. We find everywhere the same susceptibility, though not always in the same degree of forwardness or ripeness of improvement, of admitting the cultivation of these universal endowments, of opening the eyes of the mind to the more clear and luminous views which Christianity unfolds, of becoming moulded to the institutions of religion, and of civilized life : in a word, the same inward and mental nature is to be recognized in all races of men."

The differences now existing among the various human races, have been gradually developed by different climatic and other influences.

How long it may have taken to develop the most extreme differences, we are now unable to determine ; but some naturalists, as Burdach, Wilbrand, and A. Wagner, are of the opinion—that mankind in its youth was more pliable and inclined to the formation of different races, than now-a-days,—after the peculiarities of the various races, inherited through a long series of ancestors, have become more fixed.

7. WAS ADAM THE FIRST MAN ?

In 1655 Isaac de la Peyrere advanced the theory that "Adam was the progenitor of the Jewish race only, and it is only of him and his race that the Bible is designed to supply the history. Other races existed on

Earth before that of Adam; but of them the Bible contains no record, nor did the Mosaic law regard them, or impose any obligation upon them. It was only under the Gospel that they began to be comprehended in the law, which through Christ was given to all the human races of the Earth." [1]

Strange as this theory may seem to nearly all believers in the Bible, we find it defended by even such a well-known American scientist as Prof. Alexander Winchell,[2] of the University of Michigan.

Let us examine briefly the principal arguments which seem to Prof. Winchell to favor this theory.

In the first place, we know from Egyptian monuments, that—"as early as the twelfth dynasty, the Egyptians, recognized four races—the red, the yellow, the black, and the white." This twelfth dynasty ruled in Egypt about 1643, according to Strong, or 2300 years before Christ, according to Lepsius.

Now, the Deluge occurred according to Usher 2348, according to Strong 2515, and according to Poole 3099 years before Christ. From this it would seem that but a comparatively short time after the Deluge the perfectly developed negro, as he still " is to-day upon the banks of the Congo," was known to the Egyptians.

Prof. Winchell thinks it impossible that the negro-type could have been developed within so short a time after the Deluge. Hence he concludes that the negroes are no descendants of Noah—their ancestors must have existed before the Deluge and survived the same.

---

1. See Pre-Adamites, in Chambers' Encyclopedia.
2. Pre-Adamites; or a Demonstration of the Existence of Men before Adam, Chicago, 1881.

. Assuming, then, that the negroes are no descendants of Noah, could they not be descendants of Adam who, according to the estimate of some *savants*, lived about 2000 years before the Deluge ?—Would not that time have sufficed for the perfect development of the negro-type ?

Prof. Winchell thinks—not; for the following reasons: From Egpptian monuments we know that nearly 4000 years ago the negro was as completely a negro as he is still to-day ; no change can be detected ; and we have no right to assume that the negro changed only for a short time—and then permanently retained his characteristics.

Moreover, we know from Paleontology that numerous still living organic forms, as the crocodile, dog, ox, etc. existed at the time "when the negro is known to have been fully differentiated ; * * * they generally exhibit no more organic change during 4000 years than the negro does."

Prof. Winchell adds[3] : " Some of the unvarying lines of descent can be traced backward *beyond* forty centuries. Do we find them manifesting rapid changes during the next preceeding twenty centuries ? * * * No; 6000 years reveal no more change than 4000, so far as our means of measurement go. The lineage of the horse reaches back far beyond the accepted epoch of Adam, and he is everywhere a horse. By all analogies the negro-type must have persisted from an epoch more remote than Adam. * * * All the positive data tend tow-

---

3. Preadamites, pp. 216-8.

ard the conviction that the negro has come down to us from preadamic times; that he has always varied at a rate practically uniform, and that consequently his origin must not be sought in Noah, 4000 years back, nor in Adam, 6000 years back, but in some humble progenitor living on the Earth many thousand years before Adam."

In reply, we may say: Prof. Winchell is here trying to built a firm tower on a very poor and unstable foundation.

In the first place, neither the exact dates of the Deluge nor of the Egyptian monuments referred to are known. Prof. Rawlinson[4] observes: "Chronology is upon the Egyptian monuments almost non-existent. This is the unanimous confession of the Egyptologers. 'The evidence of the monuments' in respect of the chronology, says Mr. B. Stuart Poole, 'is neither full nor explicit.' 'Chronology,' says Baron Bunsen, 'cannot be elicited from them.' 'The greatest obstacle,' says M. Mariette, 'to the establishment of a regular Egyptian chronology is the circumstance that the Egyptians themselves *never had any chronology at all.*'"

Secondly, not much more reliable than the Egyptian is the biblical chronology which some learned biblicists have founded on disconnected and somewhat unsettled data given in the Bible. As observed in a former article, we have no assurance that copyists and translators have not made numerous mistakes in transcribing the numbers of years. The chronological data of the Vulgate and of the still existing Hebrew text differ con-

---

4. The Origin of Nations, New York, 1881, p. 6.

siderably from those of the famous Septuagint. "In the geneological tables of the fifth and the eleventh chapters of Genesis," as Dr. Reusch[5] remarks, "the dates of the Greek and Samaritan texts are different from those of the Hebrew and Latin." The same author does not consider it improbable that the original geneological tables have been considerably abridged by copyists.

Dr. Bernhard Schaefer[6] asserts : "It is impossible to built up a complete and absolutely reliable system of chronology from the biblical data, because of many of them we do not possess any more the original text; moreover, the Bible neither intends to give nor gives us a chronological system."

The learned Jesuit Bellynk[7] declares : "*The Bible contains no chronology.* The geneologies of the Holy Book, from which the data are derived, are incomplete. How many years are wanting in this broken chain, one cannot tell. Science may postpone the Deluge as many centuries as it finds necessary."

From all this Prof. Winchell will see that we are by no means obliged to infer—from Egyptian and biblical chronology—that the perfectly developed negro-type was known to the Egyptians, perhaps, about three or four centuries after the Deluge. For all we know, the Egyptian monuments which show the most ancient traces of the negro, may have been erected by far more than a thousand years later than the real date of the Deluge.

---

5. Bibel und Natur, Freiburg, 1866, p. 435.
6. Bibel und Wissenschaft, Muenster, 1881, p. 105.
7. Quoted by Dr. Carl Guettler: Naturforschung und Bibel, 1877, p. 316.

Now, could the negro-type have been perfectly developed within—let us say, a thousand years ?

Let it not be forgotten that Prof. Winchell's so-called negro-type is something rather indefinite.   Pritchard[8] observes that there is perhaps not a single tribe which has all the characteristics in their full perfection, that are usually ascribed to negroes.   Some tribes are perfectly black, but they have not the physiognomy of other negroes, but rather of Asiatics or Europeans.

The negro-type, then, is hardly anything so fixed and unchangeable as Prof. Winchell seems to assume.   If he can quote some authorities in favor of his views, others could be quoted that are of contrary opinion. Dr. Reusch[9] and Cardinal Wiseman[10] have mentioned a great number of authorities who seem to hold that the various types of mankind could have been developed in a comparatively short time.

A conclusive proof that there was not time enough for the negro-type to develop fully between the times of the Deluge and the Egyptian monuments referred to—is still wanting,

But let us assume Science should ever be able to show such a proof—what then ?   As suggested in a preceding article,—then we might adopt the theory[11] that besides the family of Noah, some other families, descendants of Adam, who lived beyond the reach of the Deluge, perhaps in Central Africa and beyond the Himalaya

8. Quoted by Dr. Reusch, Bibel und Natur, p. 405.

9. L. c., pp. 380-425.

10. Lectures on the Connection between Science and Revealed Religion. Lectures III and IV.

11. See Dr. Carl Guettler : Naturforschung und Bibel, p. 275.

Mountains,—survived the catastrophy which swept away all the rest of mankind known to the inspired writer. But, as yet, we are not compelled to adopt this theory. Moreover, some have suggested that perhaps the wife of one of the sons of Noah was a negress. This assumption would, no doubt, to a great extent explain the early appearance of negroes, soon after the Deluge.

Prof. Winchell goes still further; he holds that the negro-type dates back to a time anterior even to Adam. He appeals to Paleontology in favor of his view.

But although "the lineage of the horse reaches back far beyond the accepted epoch of Adam"—this is yet no proof that "the negro-type *must* have persisted from an epoch more remote than Adam,"—although the horse is everywhere a horse, and the negro a negro.

Arguments—from analogy—are often poor proofs, as also the following will show.

Prof. Winchell observes:[12] "Those who hold that the White race, the consumate flower of the tree, has served as the root from which all inferior races have ramified, may select their own method of rearing a tree with its roots in the air and its blossoms in the ground." We deny that the White race —or Adam—was the consumate *flower* of a tree; we maintain that Adam was a perfect *tree*,—not only a product of one. He came directly from the hands of the Creator, endowed and adorned with the highest possible physical and supernatural beauty and perfection; he was *created* to "the image and likeness"

---

12. Pre-Adamites, p. 297.

of God. The inferior races that have descended from Adam, have gradually deteriorated; as, for instance, also apple-trees and many other trees are well known to deteriorate, when they are propagated by seeds.

The claim that the Bible favors the theory of the existence of pre-Adamites, seems somewhat bold, in consideration of the fact that for thousands of years before Peyrere neither Jews nor Christians were led by the Bible to suspect the probability of such a theory.

The biblical passages referred to do not imply the existence of people before Adam, as J. Perrone [13] shows satisfactorily; and some biblical passages plainly declare or imply that *all* men are descendants of Adam, that *all* men have sinned in Adam, and that *therefore* Christ died for *all* men.

IX. MAN AND THE INVISIBLE WORLD.

MAN was created by God, not merely to be the crown of the visible creation, but also to be the connecting link between the material and spiritual worlds.

As to his body, man belongs to the material, as to his immortal soul, to the spiritual world.

Therefore the human soul instinctively longs for something higher, more sublime than the material world can give; the soul of man longs for God according to Whose image and likeness it has been created.

---

13. Praelectiones Theologicae, Ratisbonnae, 1854, vol. V., pp. 96-114.

St. Augustine expressed this truth in the words: "Thou, O God, hast created us for Thee, and our heart is unquiet, until it will rest in Thee."

This explains why religious sentiments are so deeply implanted in the hearts of all men. Plutarch (Adversus Coloten) could say : "If you travel through countries, you may find cities without walls, literature, laws, riches, and money, without gymnasia and theatres ; but no one ever saw a city without temples and deities, without prayers."

So deeply are religious sentiments implanted in man, that if he has lost the knowledge of the One True God, he will rather worship false gods of wood, stone, or brass, than not give vent to his religious sentiments at all.

The Bible informs us, that God Himself instructed the first parents in Paradise, how they should worship Him by the sacrifice of obedience. In case of their obedience, God promised them blissful immortality ; but, in case of their disobedience, death. The Bible relates, that the first parents disobeyed God, and thereby brought death upon themselves and all their descendents.

What have modern scientists to say on this point ?

Mr. John W. Draper [1] triumphantly observes : "The doctrine declared to be orthodox by ecclesiastical authority is overthrown by unquestionable discoveries of Modern Science. Long before a human being had appeared upon Earth, millions of individuals—nay, more,

---

1. History of the Conflict between Religion and Science, p. 57.

thousands of species and even genera—had died ; those which remain with us are an insignificant fraction of the vast hosts that have passed away."

Had Mr. Draper and other infidels, who raise this objection, carefully studied the Bible, they would have learned that the Holy Book teaches only in reference to *mankind* that death came upon all in consequence of sin.

Man alone, the crown of the visible creation, the vicegerent of God, was to be immortal, if he remained faithful to his Creator ; but of plants and animals that were created only for the use or benefit of man, Divine Revelation does not teach that they, too, had been endowed with immortality before the first sin of man.

Hence, Modern Science has, by the discovery of fossil remains of animals that lived long before man could exist, discovered nothing that ought to startle any intelligent Christian.

As to other subjects concerning Man's relations to the Invisible World, as miracles, prophecies, angels, etc., Modern Science, for obvious reasons, has discovered *nothing new* that deserves to be mentioned ; for the common, stale objections against miracles, etc. have been refuted centuries ago by Christian theologians and philosophers.

X. CONCLUSION.

HAVING carefully compared the principal Modern Scientific Views which in some way or another come in contact with Christian Doctrines, the reader will, I hope, have become convinced of the following truths :

First, although Modern Science has met with remarkable successes in investigating various *phenomena* of Nature, yet as to their most remote mysterious *causes* our scientists know substantially no more than Aristotle and other philosophers did know thousands of years ago. It is no exaggeration to say with Prof. A. P. Peabody[1] of Harvard College : "Six thousand years of research have failed to reveal in matter inherent powers that produce motion, organization, growth, transformation. We talk, indeed, of gravitation, caloric, electricity, magnetism, (etc., etc.) as if we knew what they are ; yet these are but euphemisms for our ignorance,—fence-words set up at the outermost limit of our knowledge."

Secondly, those truths or facts which Modern Science has discovered, are in perfect harmony with the truths of Divine Revelation, or the Doctrines of Christianity. Although, now and then, *at first sight*, this does not seem to be the case, yet *thorough investigation* invariably reveals the harmonies existing between the truths which God teaches both in the book of Nature and in the book of supernatural Revelation. The words of Francis Bacon will always remain true : "A little philosophy inclineth man's mind to Atheism, but depth in philosophy bringeth

---

1. Christianity the Religion of Nature, Boston, 1864, p. 64.

men's minds about to Religion." It is never true Science that is dangerous to Religion, but only such superficial knowledge as now-a-days puffs up many a one who has not yet learned enough to see how little he really knows. Such people swell the ranks of Modern Infidelity, because they readily venture to decide on even the most difficult questions of Science or Religion, before they are able to examine them fully. It is not true, thorough Science—but the want of it—that produces scoffing unbelievers.

Whatever Science may discover, in the midst of human theories that continually succeed each other and change like the billows of the ocean,—one thing is certain—the Rock of God's Word will never be shaken.

"THE TRUTH OF THE LORD REMAINETH FOREVER." Ps. 116, 1.